D0189312

MIND OVER
MATH

MIND OVER

STANLEY KOGELMAN, M.S.W., Ph.D.
and
JOSEPH WARREN, Ph.D.

McGraw-Hill Book Company
New York St. Louis San Francisco Bogotá Düsseldorf Madrid
Mexico Montreal Panama Paris São Paulo Tokyo Toronto

Reprinted by arrangement with The Dial Press
First McGraw-Hill Paperback edition, 1979

3456789OFGRFGR 83210

Library of Congress Cataloging in Publication Data

Kogelman, Stanley.
Mind over math.

1. Mathematics—Study and teaching—Psychological aspects. I. Warren, Joseph, 1936– joint author. II. Title.
[QA11.K74 1979] 510'.7 79-15176
ISBN 0-07-035281-X

To
Elaine Sorel
for her warmth, friendship,
and inspiration

contents

PART THREE
mind over MATH

PART FOUR
doing MATH

PART FIVE
after MATH

PART SIX

supplemental MATH

preface

Although we are both mathematicians, we came to understand people's fear and dislike of mathematics in different ways. After several years of doing mathematical research and teaching in college, Stan found his interests changing and moving toward psychology. He decided to enter a master's program in clinical social work at Smith College where he discovered that his fellow students had no difficulty with any of the psychology courses but dreaded the required statistics course.

The intensity of their feelings was striking and Stan began to wonder if math difficulties might have emotional roots. He researched the causes of math anxiety in his thesis, which was completed in the summer of 1975.

In early 1976, at an open house at Wesleyan University in Middletown, Connecticut, Stan met Joseph Warren. A math clinic had just been established at Wesleyan with the support of a federal grant under the leadership of Ms. Sheila Tobias, Associate Provost of Wesleyan, and Professor Robert Rosenbaum, a mathematician. Stan later became a consultant to the math clinic.

Joe's interest in math anxiety developed from his experience in teaching and working with people individually. He had observed that, for many, no amount of patient tutoring enabled them to learn math. He had personally experienced difficulty with math in the early grades at times of stress that had nothing to do with math. He realized math problems could be more emotional than intellectual.

In the spring of 1975, Joe started a program for individuals called "Math Therapy." He found that people were able to learn more math more quickly if they discussed their feelings about the subject with him.

We both realized there was a general public need for programs to help larger groups of people overcome math anxiety. We decided to start such a program in New York City and felt a group approach would be most effective. Ms. Tobias introduced us to Elaine Sorel, who had a background of working with artists, writers, and directors. She specialized in developing, presenting, and promoting new ideas.

Together, the three of us (Elaine, Stan, and Joe) developed "Mind Over Math," which was founded by Stan and Joe in June, 1976 as a consulting service offering programs aimed at reducing math anxiety for schools, colleges, corporations, and groups of individuals.

This book is based primarily on our experience in leading groups to help people overcome math anxiety. Chapter One is based, in part, on Stan's research on the causes of math anxiety. The initial writing was a joint project, the final writing being done by Stan.

The inspiration for this book came from Elaine Sorel, who has extraordinary vision and intuition. Her creativity, support, insight, and enthusiasm has furthered the growth and the development of Mind Over Math.

Our warm thanks go to all those who have so openly shared

their feelings and experiences about math.

We are indebted to Mimi Clifford, Noreen Goldman, and Maryann MacBride for their caring support and invaluable suggestions throughout the course of this work. Stan especially wants to thank his daughter Laura for her patience, love, and understanding.

Stanley Kogelman
Joseph Warren
March, 1978

introduction

Many people react to math so strongly that their ability to memorize, concentrate, and pay attention is effectively inhibited. This makes learning impossible. It also makes testing math ability impossible, because often all that can be assessed is the test-taker's math anxiety.

We cannot believe that someone who performs well in other disciplines would have no ability in math. This is not to say that each person should be able to perform equally well in all areas. But we do feel that the difference in ability should not be as wide as it often is.

We found that math anxiety can be overcome and performance improved by participation in a series of five weekly Mind Over Math workshops focused on approaching math rather than on solving problems.

The atmosphere of the workshops suggests a living room more than a classroom. There is no blackboard, and there are no tests and no homework. Coffee is always ready, and cookies, cheese, and crackers are laid out. In the first session, a group of ten people sit in a circle and recall their experiences with math: specific teachers; early difficulties; present feelings.

Like the Mind Over Math workshops, this book will help overcome math anxiety gradually. The first five parts follow the themes of the workshops. Each successive chapter lays the groundwork for decreasing anxiety. Material that might be anxiety provoking if looked at immediately will be less upsetting later on. It is important, therefore, to read the book in sequence. At times readers may feel like skipping certain parts. We urge them to resist this temptation.

In Part One, as in the first workshop session, no math is discussed. The focus is on the nature, causes, and effects of the math anxiety experienced by so many people.

Part Two, "Math Dynamics," will give readers deeper insight into their interactions with math. This insight will make it possible to *break the cycle in which anxiety controls people's approaches to math.*

Part Three, "Mind Over Math," reveals that doing math is not much different from doing anything else. *The same skills that bring success in other areas of life can be utilized in learning and doing math.*

Part Four, "Doing Math," offers practice and reinforcement, as well as practical suggestions.

Part Five, "After Math," describes how the techniques used to overcome math anxiety can be carried over to other things that people feel they "just can't do." It is an interesting fact that, in the course of reducing math anxiety, people find they can suddenly do seemingly unrelated things like change fuses, use a reflex camera, follow instructions, read legal documents, and even play tennis. The removal of an apparent deficiency in thinking leads to improved self-confidence and self-image. These changes in attitude show up in increased assertiveness in both daily and job-related activities.

Part Six, "Supplemental Math," provides an overview of math topics for further study, after the fear of math has been overcome.

PART ONE

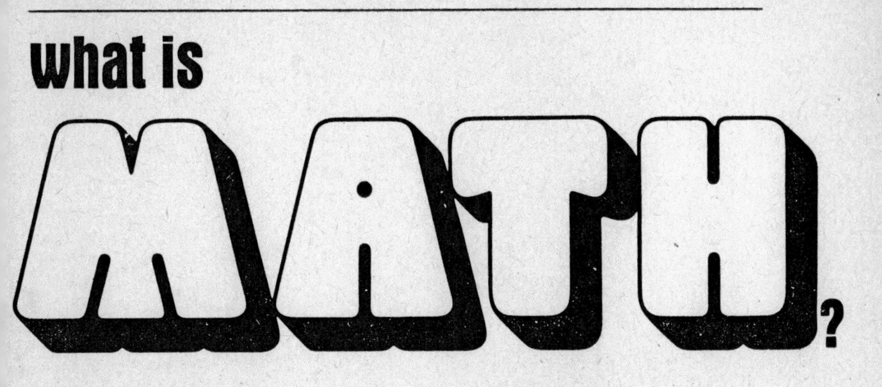

CHAPTER ONE

what is math anxiety?

After looking forward to it all week, we had finally arrived at Karen's party. There were just the right number of people. Her parties were always relaxed and fun and we knew we'd make new and interesting acquaintances since so many of her friends were artists, writers, actors, producers, and executives. Music was playing, but not too loudly, and the food looked fantastic.

We told Karen we wanted to meet Betty and Susan, two successful business executives. Karen felt sure they'd be interested in meeting us. But when she introduced us to them by saying we were mathematicians, they instantly turned off. "Why didn't you tell me that there were going to be mathematicians here?" Susan said accusingly to Karen. They looked us up and down and visibly began to withdraw. Within a few moments they had excused themselves to get some drinks. We saw little of them for the rest of the evening.

It is disappointing when people we are attracted to avoid us just because we are mathematicians. All the mathematicians we know have similar experiences with most people they meet. Joan, a workshop student of ours, explained that she would never expect to be able to talk to a mathematician about anything. "They are a different species, it would be like talking to a grizzly bear. There would be no connection between my life and theirs. I would expect a mathematician to be slightly contemptuous of me."

Karen is a close friend and colleague, but not a mathematician. She recalled that she had been very anxious about our first meeting:

> I couldn't imagine having one mathematician in my living room, let alone two. What would we say after hello? I expected you to be unattractive, cold, and to have absolutely no sense of humor. I'd be funny and silly and you wouldn't get into that at all. How could I possibly be serious for more than twenty minutes? Basically, you would be boring!
>
> I thought you would be wearing little old men's shoes with laces. Your pants would be too big, too wide, dark and baggy. You would be wearing a white shirt, be out of date, and just not know how to put it all together. It wouldn't even occur to you that the pants were too big—like the absent-minded professor who goes into the store and the guy says I don't have a 34, only a 42, and you say, "I'll take it!" You would be out of touch with that part of the world and it would have no importance to you. The only world you would relate to with any intensity would be mathematics and physics. You certainly wouldn't know who the Rolling Stones were. And you would carry a very ugly briefcase —worn, rotten-looking, and old-fashioned.
>
> I can't describe the surprise I felt when I opened the door and saw two attractive young men. You wore jeans and had long hair coiffed in a very stylish way. The physical aspect really threw me. And you had a sense of humor . . . you were responsive to me.
>
> I expected there to be huge differences between us because we

wouldn't be able to speak the same language. I felt that you were impressed with me, but I was aware of the fact that I was a dummy when it came to math. I wanted to keep it from you and felt that my image would be altered if you were to find out that I didn't know math. Until you mentioned it, I never thought about being anxious about math . . . just about not knowing it and being stupid. I don't think I ever heard the word anxiety applied to an inability to do math.

Now I think I would be more open to meeting a mathematician. Still, I think that underneath those dungarees there is a pair of baggy pants.

Karen's graphic description captures the feelings many people have about mathematicians. They are expected to be different than other people—cold, withdrawn, introverted, and more comfortable relating to figures than to people.

What can you talk to a mathematician about except numbers and theorems? Although some mathematicians do fit the stereotypes, most do not. Research on the personalities of mathematicians has not supported any of the expected stereotypes.

Feelings most people have about mathematicians are related to their feelings about math. In order to deal with math more successfully it is necessary to come to a realistic view of mathematics and mathematicians. Even when people find we are like "normal" people, that is, we talk, drive cars, play sports, and have sex, they are inclined to say we must be the exceptions. Almost all mathematicians have difficulties with math at times and have experienced anxiety in doing it. Very few mathematicians are geniuses.

Many men and women from students to corporation presidents suffer from math anxiety, an intense emotional reaction to math based on past experiences. This reaction guides and controls their approach to math to such an extent that doing math becomes extraordinarily difficult if not impossible. They come to regard themselves as having a learning disability in this area.

But difficulty with math does not stem from a learning disability. It is really a question of attitude rather than aptitude, and there are ways that your own anxiety and negative feelings about math can be overcome.

Perhaps your own goal is only to be able to do everyday math—to check the bill in a restaurant or balance a checkbook. You may want to take a math course or you may have to bone up for the business boards or the graduate record exam in order to go to graduate school. It may be necessary to read graphs and have facility with figures in order to advance on your job. Or, you may simply want to "get over a mental block." Most important, you may want to be sure important decisions concerning jobs, professions, or college majors are based on deep personal aspirations rather than fear and avoidance of math.

Some of the effects of math avoidance have been documented in a study at Berkeley by sociologist Lucy Sells. She found that of freshmen admitted in 1972, forty-three percent of the males and ninety-two percent of the females had not taken four years of high school math. But this much math was a requirement of fifteen of the twenty majors at Berkeley. In effect, those students had limited themselves to only twenty-five percent of the possible majors.

This is a book about widening options through making the best use of all your abilities and increasing awareness of talents you never realized you had. The neglect of the individual personality which often occurs in traditional learning situations leads to feelings of frustration, discouragement, and general math avoidance. *Mind Over Math* provides a new approach to learning and doing math based on getting to know yourself and your individual learning personality. It is not a math text.

Participants in our Mind Over Math workshops range in age from fifteen to sixty-five. Some are high school dropouts and some have Ph.D.'s in the humanities. They come from a wide range of cultural and ethnic backgrounds, and two-thirds are women. The differences in their backgrounds contrast sharply with the similarities in their feelings and experiences with math. They often express hatred of math and intense emotional and physical responses to it. For example:

Joan: If you want to see me panic, all you have to say is, "If two men can dig a ditch in one hour . . ." My heart starts beating even when joking about it.

Barbara: An asbestos curtain drops in front of my eyes.

Bob: When I see figures I immediately feel "I can't do it. TAKE THEM AWAY!"

Jane: I know I won't understand. I look at a page of symbols and just want to leave.

Jeff: Numbers drive me up a wall!

Mary: I can't stand to add my check book. I avoid it. It's crazy. I get nauseous.

Hal: I just panic, block, freeze up.

David: Being in a math class is like being in the ring with Muhammed Ali . . . incessantly peppered in the face and totally helpless.

The feelings expressed are strong and extremely graphic. We might begin to wonder what it is about math that can evoke such emotions.

It is not abnormal to have anxiety. Most people experience some situations as so uncomfortable that they get tense and simply cannot operate normally. Some suffer from writing blocks or language blocks. Others seem to have two left feet when dancing, or feel absolutely spastic in sports. A few find they get flustered over a recipe and most fear getting tongue-tied before a large audience. Whatever the area, the simplest and most common means of eliminating anxiety is by avoiding the provoking situation.

It is not surprising that someone would want to avoid the painful feelings math evokes. Doing math in the presence of intense anxiety is all but impossible. Once panic begins to take hold, normal functioning is impaired and the skills necessary for learning and performing become inaccessible. It is then impossible to work up to capacity or even discover what these capacities are. *But since this is an emotional, not intellectual inhibition, it can be overcome.*

Solving a math problem or learning from a book and teacher requires concentration and clear thinking. Information must be taken in by paying close attention to what is written or said. It is also necessary to remember what has been learned so that the next topic can be understood. Unfortunately, anxiety and panic seriously interfere with memory, attention, and concentration.

It works like this. If someone is swimming and suddenly finds he is out too far he may either panic or calmly take in the situation. If calm, he may consider the distance to shore and the strength of the current and notice if there is a lifeguard or people nearby. Then he will decide what to do. He may choose to swim to shore slowly or may feel the current is too strong. He may call to someone nearby for help or wave his hands to attract attention.

In other words, if he doesn't panic, he is able to devise a strategy to get out of danger. When panic occurs, however, he cannot think clearly enough to be able to take in such details of the situation as the distance to shore, current, and availability of help. Even when help arrives, there is a tendency to panic while being saved. Fear of drowning and helplessness are the only things in mind.

When panic occurs it is tempting to try to talk yourself out of it by rationalizing or to suppress it. But these approaches rarely work. They do not make math anxiety go away or lift that "asbestos curtain." Later in this book, constructive ways of reducing intense reactions to math will be developed.

The first step in overcoming math anxiety is to accept your feelings and realize they are not unusual. In fact, more people have these feelings than don't. Even as mathematicians, we find many math-related situations in which we experience waves of anxiety. When Joe is unexpectedly asked to do math in front of other people he tenses up and gets a dull feeling in his head. He thinks:

> I don't want to do it. My mind is not going to work. If I can't avoid this, I'm going to have to go into another room to do it.

Stan has a similar reaction. He recalls panicking on math tests and thinking:

> They're going to ask exactly what I don't know. I'm going to fail. They will look at my paper and decide I'm too dumb to be a math major.

All mathematicians we have known recall having had feelings similar to ours that interfered with their ability to do math. The way that intense emotions typically affect memory, attention, and concentration are best expressed by people who have experienced math anxiety:

> Barbara: Sometimes when numbers come at me fast and hard, I can't distinguish one from the other and I check out.

> Mary: There was something about math, about numbers. My thinking would get clouded. I can memorize other things but when it comes to math I couldn't memorize if I didn't understand it.

> Alice: If someone were tutoring me in math, I would get so upset that I could hardly hear them explaining. I would block, as if there were an internal mechanism going in my mind which said, "I can't do it, I can't do it, I can't do it." I was no longer hearing. It felt like an intellectual weakness of the brain tissue.

> Peggy: There is difficulty absorbing it and concentrating on it. My mind wanders.

> Nancy: If you tell me the history of something I'll remember it in one second. But when it comes to numbers or descrip-

tions of how to do something, like a manual, the shade comes down. I forget a lot. It just doesn't stick.

David: In math class the instructor was speaking clearly, but I just couldn't take it in. I couldn't hear his words . . . though I sat in the first row.

Hal: Even when I can do math, I write the wrong answers. The other day I was writing a check for $89 and when I got to the bank, they said I couldn't cash it. It said $892.

Clearly these people are not functioning normally when it comes to math. Alice and David don't have hearing problems, but they can't take in what is being said in math. Mary and Nancy have excellent memories and always performed well academically, but couldn't remember when it came to numbers. Barbara, Peggy, and Hal are all successful in their work, but have great difficulty concentrating on math. They may all feel they don't have the ability to do math. But no one could possibly do it when such strong reactions are present. Panic blocks the normal thought processes. Not even a genius can work when extremely agitated and upset.

It is helpful to understand why attempts at suppressing adverse reactions to math don't work. Think of it this way. You only have a limited amount of emotional energy available. A lot of this energy goes into anxiety and panic. Then if you give yourself a pep talk and say, "Stop it! Calm down! Don't get so worked up!" a great deal more energy goes into calming yourself. The combination of these efforts is so emotionally draining that there is little energy left for work. It is similar to what occurs in the following isometric exercise: Stand up and press the palms of your left and right hands together. You will find that as you press harder and harder, your muscles are stressed but nothing moves and no work is done. There is just a muscle flex. This is great for building muscles but this kind of brain flex accomplishes absolutely nothing. It is just frustrating and discouraging.

It is better to understand the origins of adverse reactions to math than to attempt to suppress them. Recognition of the influence of the past diminishes its effects in the present. Many people remember their first negative experiences with math with such clarity and emotion that it seems as if it were just yesterday, even if it really was twenty, thirty, or even fifty years ago. They even remember how a teacher looked, smelled, or dressed.

Joan: In algebra I was given this little immigrant teacher who I don't think had any shirts. He came to school with a blue coat and a blue blazer with a big safety pin in front, spoke in a very heavy accent and didn't understand us. I certainly didn't understand him.

Barbara: I remember my high school geometry teacher. He was a kindly man in a rumpled gray suit. He was gray all over. He had a redheaded daughter who was embarrassed because her father taught in the school. And there was a skinny little birdlike boy in the class who knew everything and would dart up to the blackboard all the time.

Negative math experiences most frequently occur between the seventh and tenth grades. Often people recall having been comfortable or even excelling in math until then. However, some people don't remember ever having been comfortable with math and may have had painful experiences in elementary school. Memories similar to what Hal and Bob movingly describe are quite common:

Hal: I was always one step behind the class. They used to say things like, "When you have the answer, stand up!" and I would be the only one sitting down. My best friend would be standing and would say, "Come on, you can do it!" Meanwhile, the teacher would be looking down and saying, "Who isn't standing up?"

This would go on day in and day out. It got to the point where your best friend didn't want to be seen with you. I would be the only kid in the class who couldn't do it in a logical fashion.

Bob: I can't tell the exact time when I was aware that I really couldn't do math. It probably stems back to my elementary school days when I went to parochial school. They were very harsh and it was a very strict atmosphere. If you didn't get it, they berated you and made you stay in your seat for hours until you got it.

There were many times when I cried, "I've got to go home, please!" And they said, "No! You have to do that! Work it out!" They made you sit there and do it until either you did it, or you collapsed on the desk from crying so hard. Then finally they let you go.

Hal and Bob both felt humiliated at being singled out and experienced overwhelming pressure to get an answer. But getting one would have been almost impossible since no one could be expected to think clearly under those circumstances. The more they tried, the worse it got. The intensity of the feelings expressed in relating these incidents indicates how traumatic they must have been.

Others have described the traumatic experience of being called to the blackboard and forced to stay there until they got an answer, or of being physically punished if they couldn't do a problem or failed to do it fast enough. Obviously, learning math in an insensitive or punitive environment contributes greatly to present feelings. For Hal and Bob, each contact with math brings back the painful feelings and anger of the past.

For some there was no direct humiliation but rather the indirect message that went with being ignored:

> Gloria: I remember my seventh grade teacher. She didn't care. She couldn't be bothered with me because I was too slow. I just didn't understand what she was trying to say. She would pass over me. Sometimes I would go after school and ask her for help. But she couldn't help me. She would just dismiss me and I was left in a fog.

Gloria was very discouraged and frustrated by her teacher's refusal to help her and answer questions. She blamed herself and felt she was just "too slow."

The questions students have about math should be encouraged; frequently they are very astute and difficult to answer. Unfortunately, students often become intimidated and feel that if they were smarter they wouldn't have so many questions. This is not the case.

The progress of mathematics has been uneven. Major innovations have usually developed over decades or centuries because of the difficulty involved in conceptualizing new ideas. The points at which people feel confused in math are usually in the areas that were difficult historically. The questions they ask could very well be the same ones asked by the mathematicians of the past. It would be helpful if the students knew how good their questions really were.

Negative experiences with math do not always relate to a teacher. Being tutored by a parent or sibling can be very stressful and may leave one feeling frustrated, inadequate, and angry. Competition with siblings and comparisons with an older brother or sister can be immobilizing.

> Barbara: They skipped me in the second grade and in the third grade there were numbers which I had not seen before. I was now in the same grade as my sister, who was only a year older than I. That was very embarrassing for her. They even made the hideous error of putting us in the same class in the fifth grade.

> Pat: In the fourth grade I was out of school for a few months and my mother was afraid that I had missed something. She grilled me in fractions and multiplication tables. She would get angry and impatient when I couldn't do it. I remember feeling I was incapable of doing arithmetic. I thought my mother was going to hate me for it.

A close relative seldom has the required patience and objectivity to teach you something difficult and frustrating. It's like the old sexist joke about how many divorces result from a husband trying to teach his wife how to drive. When a relative becomes impatient and angry at a child it is even more hurtful than when a teacher does the same thing.

A prolonged absence from school or a move to a new one can have a particularly adverse effect on a child's feelings about math. Since mathematics builds upon itself, missing an idea or concept can be very confusing. Until the missing topic is learned, there remains a gnawing feeling that something basic should be understood but is not.

As your own anxiety about math decreases, it will become easier to isolate exactly what you do not know. Then it will be possible to learn the missing topics from a book or by asking someone who is able to answer questions clearly and patiently. The gaps are never as large as they appear. An adult who has overcome math anxiety can usually master arithmetic in a short time and learn more advanced subjects in about half the time required in high school.

It is clear that the kinds of experiences just described contribute to negative feelings and attitudes toward math. The vividness of the memories and the intensity of feelings indicate how significantly they may affect the present approach to math. New experiences become particularly traumatic when they reinforce and accentuate previous adverse feelings about math. The result can be a long-term hatred and avoidance of mathematics.

Math anxiety does not have a single cause but is the result of different factors working in resonance with bad experiences. One of these factors is the perception of math as a masculine pursuit, which may begin as early as elementary school. Young children are always seeking role models and in elementary school they find that most teachers are women. Teachers often dislike teaching math and prefer to teach reading and other subjects. They may consciously or unconsciously transmit their feelings about math to their students. If a girl identifies with her teacher it is possible for her to begin to adopt her teacher's attitude:

> Mary: Math was fun in the beginning of school. In the fourth grade the teacher announced to the class that she hated math and was going to have a special teacher come in to teach it.
>
> I loved my teacher. She was terrific. She was very sweet and kind and understanding. There was something about her being a woman and being interested in both English and history. I remember thinking during that year that I didn't like those numbers either. I would rather not have to deal with them. I wished they would go away.

The influence of Mary's fourth grade teacher is clear. But teachers by themselves are not to blame. There is a strong message given by society that it is men, not women who are expected to handle money and finances and prepare themselves for careers. These ideas may be changing but they are still very much entrenched in our culture. It is difficult to escape their influence. This is clearly expressed by Eleanor:

> I learned from childhood that, in a restaurant, the man checks the check. I'm sure that part is a cultural thing—men have the careers and are supposed to be the smartest. Women stay home and raise children.
>
> There was even some satisfaction in not being able to do math, because it was feminine and girlish not to do it. It's okay if you don't do well in math and science, it's the boys that have to.

If facility with math is viewed as masculine, a girl may feel that excelling in it conflicts with her self-image. Rejection and avoidance of math then have the positive qualities of asserting a girl's feminine identity. The negative aspect of math avoidance is that most careers become inaccessible because they require math at some point. This leaves women with few options besides nursing, teaching, and the humanities.

Although attitudes toward math develop early, it is usually junior high school and high school that mark a real turning point. This coincides with the adolescent period during which boys and girls are struggling with independence and trying to establish their identities as men and women. At the same time math becomes increasingly abstract as it changes from arithmetic to algebra and then geometry and trigonometry. Boys are encouraged to study math so they can become engineers and scientists and they are expected to think logically, not emotionally.

Ability to do math is often mistakenly viewed as the only true indicator of high intelligence. It can be socially undesirable for a girl to appear too intelligent, especially if it is reflected in her math performance. She may even feel the danger of social ostracism if she does better than the boys. Being bright and mathematically talented may conflict with being feminine and emotional.

Noreen: I had been quite good at math until the eighth grade. Then I became aware of boys. I went absolutely pot over tea kettle. This was a man's, no a boy's, subject. From then on it was all the way down.

Susan: At one point I was a tomboy and didn't know whether I was a boy or a girl. I decided to be a girl. In college I changed my major from engineering to journalism because engineering was for men. If I became an engineer, people would laugh at me. My image of female mathematicians is that they wear white coats and may be pretty but it is all covered up. They are hidden beauties.

Not every young woman is going to reject math and feel she can't be feminine and mathematical at the same time. But if she is struggling with her identity, sees math as masculine, and gets reinforcement from parents, teachers, peers, and society, then math becomes something to be rejected.

Girls are often directly and indirectly supported in their resistance to math. When they wish to drop math in high school, they meet with little opposition, while boys are encouraged to take as much math as possible, so that career options remain open. As Karen recalls:

> In the ninth grade, obviously by some mistake, I was put into an algebra class. After about three weeks I said this is not for me. It was the first time I realized I could talk my way out of something. I told my advisor I didn't need this, and wasn't going to go to college. I just asked him to let me out of the class. After about three weeks he did.

Lack of positive reinforcement for doing math can be taken as permission not to do it. Parents, teachers, and counselors often fail to stress the importance of a young woman's taking as much math as possible, because of the traditional view that she is to be a wife and mother and not a career person.

So for women, math anxiety and avoidance can result from a combination of factors which include sex role stereotypes, identification with female teachers, and active and passive reinforcement by peers, teachers, counselors and society.

Attitudes toward math can also be shaped by the perception that it is a rigid subject consisting primarily of a set of rules to be obeyed. This authoritarian image can cause math to become associated with feelings toward authority in general. During the course of growing up, children experience restrictions of their freedom from parents, teachers, and society. When they find they can't do what they want, they may object to these restrictions. Rebellion against authority is a normal part of adolescence.

Schools and classrooms have a built-in atmosphere of authority. Everyone has to go to school whether they want to or not. In school, pupils are told what to do. They raise their hand when they want to speak, get permission to stand, and even have to ask to go to the bathroom. This structure limits personal freedom. Added to this is a system of grades and evaluations. If math is perceived as an authoritarian subject, feelings about it may merge with general feelings about authority. Then adolescents in junior high school and high school may refuse to have anything to do with math.

Math does not easily lend itself to being taught in a group format where ideas evolve through discussion. The history of mathematical ideas and the lives of mathematicians are rarely a part of learning mathematics. The emphasis is on communication of a specific body of knowledge, and this usually is done through lectures and drills. This style may result in the feeling that math is impersonal and detached from human experience:

Barbara: I found something humiliating about inhaling all this stuff and then spitting it up and going on and never understanding it. There was something so coercive about it. Why did I have to memorize it just to pass some standard or please somebody? I could never understand it. It didn't make sense to me, so I could never take it in.

The human expression in math comes through the personal ways math is conceptualized and problems are solved. These qualities are not usually conveyed in math classes, but will be brought out in succeeding chapters. It is usually felt that the teacher is the possessor of the ideas, concepts, and methods of math and is just passing them on to the students. Many feel as Barbara, that no part of math comes from them.

Math books contribute to the rigid image by the formal style in which they are written. There is usually no talk of how people do math, but only explanations and the logical presentation of concepts and methods. This may appear inflexible and forbidding. It is as if math were something that has always been there, not something men discovered and developed. It may be felt that math is eternal truth and is to be accepted, not understood:

Dolores: I went to parochial school and my math class came right after religion. We had to accept religious doctrine as The Truth. When I got to math class, I felt that I had to do the same. I remember not wanting to learn it and finding religion easier to accept.

Questions in math books and on math tests don't have the informal quality of questions in the humanities, which require discussion of a topic, book, or idea. Problems usually have only one answer and this can be very disturbing.

Wendy: I have to use numbers a lot in my work but I find that even if I get the answer right, I'm not sure of it. I'm afraid of being wrong. I do it over two or three times because it worries me that it's not right.

Jeff: There is a difference between a lack of grace and being wrong. In math you are right or wrong. In English, you are clear or muddled.

Rosemary: I recently had to take over doing the tax returns because my husband was sick. The Internal Revenue Service sent back a statement saying, "You made an error in calculation." It was very humiliating.

In most other subjects you can be almost right even when you are wrong. For example, Stan recalls his English instructor reading a sample of poetry and asking the name of the author. Stan said, "Frost" and the instructor said, "No, but you have a good point, for his style is similar to Frost's."

Although there is a right-wrong quality and logical structure to math, it need not be presented in a rigid way. When math is demystified in the next chapter, it will become clear that intuition and creativity are essential in learning and doing math. Math is bound to be distasteful if it is experienced as rigid, judgmental, and inflexible. When this feeling is amplified by resentment of authority or a bad experience with a cold, rigid, and impersonal teacher, rejection and avoidance of math is likely.

The focus on detail and symbolism required by math can be very unpleasant, frustrating, and annoying and may be another cause for rejecting it. Many people hate details of any kind and even dislike nonmathematical things such as forms, instructions, and extensive menus. Their impulse may be to push detailed material away and avoid looking at it. This can also apply to daily experience:

> Kathy: I have a tendency not to see things in the world clearly. I have to consciously make myself aware of detail. I block out a lot of what I see.

The alternative to concentrating on detail is to get an overall sense of a subject and rely primarily on feelings and intuition. In math books, however, there is no way to get a sense of the concepts without careful reading. Each word and symbol has a specific meaning that must be understood, so reading a math book is very, very slow. In a typical one-term literature course students may be expected to read twelve to fifteen books of about three hundred pages. The same length math course would cover only two hundred pages in a single book.

When there is a very strong personal preference for concrete realities and feelings, the extensive use of symbols in math can be experienced as dehumanizing and offensive.

> Peggy: I hate everything being defined in terms of letters and numbers.

> Helen: High school math was a total mystery to me. There was all that symbolic thinking. Those awful symbols. . . .

When someone feels this way and is forced to take a math course, memorization may be the only way out. With extreme effort, it is sometimes possible to remember long enough for a test, but afterward there is almost total forgetting. With memorization, the details and symbols remain foreign, fearful, and meaningless. The concepts are never internalized and there remains a constant sense of not understanding. This leads to further insecurity and eventual giving up.

In addition, the required mastery of details in math may be experienced as rigid and authoritarian by those who have a divergent thinking style. Some people especially like to use their imagination to generate a variety of ideas and images about a subject. They may resent having to obtain a definite answer by following a formal sequence of steps. Since they don't do this in other subjects or in daily life, they may feel that math forces them to think in an unnatural way. These feelings are clearly expressed by Peggy:

> I don't like the idea of having an answer. I like things where there is no right or wrong. The specificity, the exactness, disturbs me. When being tutored, I wanted to move faster and pass over things. Anything that has a logic to it, where there are steps involved and a process that I have to know, gets difficult. I hate those damn formulas.

Although math does require a more focused, convergent approach than other subjects, imagination, emotions, and intuition play a critical role in all mathematics. Many of the feelings people have about the cold, logical, rigid qualities of math are related to firmly entrenched myths. The next chapter is devoted to examining these myths so that math can be approached with more comfort and realistic expectations.

twelve math myths

Many of the most commonly held views of math are based on myths about the subject. These myths have resulted in false impressions about how math is done. They need to be dispelled.

ONE: MEN ARE BETTER IN MATH THAN WOMEN.

Research has failed to show any difference between men and women in mathematical ability. The perception of math as a masculine domain stems from other myths about the subject. Math is seen as the epitome of cool, impersonal logic—nonintuitive and abstract. This fits with the stereotypical image of men as cool, detached, and objective.

Men are reluctant to admit they have problems so they express difficulty with math by saying, "I could do it if I tried. I just never worked hard enough." Women are often too ready to admit inadequacy and say, "I just can't do math. It's an ability I just don't have." Both may be expressing the same fears or anxieties about math. Russell Baker, a *New York Times* columnist wrote humorously about his experiences with math (September 25, 1977):

> It is barking up the wrong tree for women to conclude that mathematics is more terrifying to them than to men . . . I first started fearing math near the end of the seventh grade . . . Many desperate years later I was one of that vast multitude of men who emerge from college without a doctorate in mathematics and have been scarred for life by inability to do our income-tax returns or verify the addition on restaurant checks . . . If men get more doctorates than women it is surely not because mathematics is "male-oriented," but because males, with their powerful instinct toward machismo, are ashamed to admit that when it comes to *pi* they are chicken.

Women mathematicians are expected to be less feminine than other women. This is not true. In fact, a 1960 study at UCLA by Phillip Lambert, which measured femininity of interest patterns in terms of current cultural stereotypes showed that women math majors "were not only equal to nonmathematics majors in femininity, but significantly more feminine."

If more people saw math as an intuitive, creative human endeavor, then women and men might be expected to have equal ability and interest in mathematics.

TWO: MATH REQUIRES LOGIC, NOT INTUITION.

Few people are aware that intuition is the cornerstone of doing math and solving problems. The final product of mathematical work is completely different in appearance from the process by which the result was obtained. Einstein wrote:

> To these elementary laws (of physics) there leads no logical path, but only intuition supported by being sympathetically in touch with experience.

Newton, who invented the calculus, was unable to give the logical mathematical basis for his ideas. It took two hundred years and tremendous mathematical advances before his work could be "proved" correct.

Mathematicians always think intuitively first. The logical presentation of results follows, and may require far more work than coming up with the solution. The conception of math as strictly logical conflicts with the definition of intuition—"the act or faculty of knowing without the use of rational processes."

Math books may give the impression that math requires an especially "logical mind" because they are so ordered, precise, and logical—and because they are often impossible to follow. Even when a person can follow all the steps, there is a tendency to think, "I see what they did, but I could never have done it myself. I don't have a mathematical mind."

Math teachers usually spend considerable time preparing their lectures and working on problems. When problems are presented in class, the solutions have been revised to the point where they have become brief and logical. The scrap paper that represents the teacher's process of arriving at the solution is thrown away. Students are left with the impression that this carefully refined solution was arrived at by a magical process or divine inspiration.

In a painting you don't see the sketches, planning, and struggle that produced the finished work. Final mathematical form is the result of a process of using intuition, making false starts, and struggling until the solution is "seen." It is no more possible to sit down and solve a problem you have never seen before than it is possible to produce a work of art at one sitting by starting at the top of the canvas and working directly to the bottom.

In our society, intuition tends to be undervalued while logic and reason have the aura of religious truths. We all tend to react emotionally and intuitively first and then explain what we feel by means of logical argument. It is extremely difficult to convince anyone of anything through the use of logic alone. Logic can be used to support a point of view but in the final analysis the listener must be persuaded that the concept "feels" right.

Mathematical intuition refers to ideas about the answer or method of solution to a math problem. These ideas may not have evolved logically but rather seem to just pop into your head:

> Michele: I asked my bank manager to explain compound interest but two elements of his explanation went absolutely over my head. When I got home, I decided to pretend it was *I* who had to explain it to someone else. When I reached the first point that I had been unable to understand, I hesitated for a moment. Suddenly it became crystal clear. This must be the *intuition* we discussed yesterday.
>
> The second puzzlement was harder to unravel. I started playing with the numbers and found relationships I had not seen before. Again, in a flash, all became clear.

Everyone has mathematical intuition; they just have not learned to use or trust it. It is amazing how often the first idea you come up with turns out to be correct.

THREE: YOU MUST ALWAYS KNOW HOW YOU GOT THE ANSWER.

Getting the answer to a problem and knowing how the answer was derived are independent processes. One involves intuition, the other logic. If you are consistently right, then you know how to do the problem. There is no need to explain it.

Mathematicians are not always aware of the process by which they do math. Lightning calculators are able to do extensive computations in their heads with amazing speed and accuracy but rarely know how they get their answers.

What may appear to be guessing is really mathematical intuition at work. Demanding an explanation inhibits the use of this intuition and robs one of the pleasure of getting right answers!

> Joan: I could get the right answer in school, but the proof had to be part of the answer. You had to know how you arrived at it. That was impossible for me so I felt that if I got the right answer, it was just luck. Somehow I had arrived at it but, since I didn't know how, I had to assume a little bird told me.

Proofs of mathematical results have not always been valued. Historically, the Arabic school of mathematics emphasized getting answers while the Greek school put proofs on an equal footing with results or answers.

Examining the process by which answers are obtained is

useful in those instances where you find that consistently wrong answers keep appearing in the solution of a particular type of problem. In this case, you are concerned about where you are going wrong so you can correct yourself.

Teachers have to know methods so they can transmit them to others. Yet each person eventually comes up with his or her own methods in math, just as everyone develops a unique handwriting—even though everyone was taught the same way. This is the creative part of doing math.

FOUR: MATH IS NOT CREATIVE.

Creativity is as central to mathematics as it is to art, literature, and music. Mathematical theorems may seem to be "just thought up," but they actually represent the end result of the same creative process that leads to a painting or symphony.

The act of creation involves diametrical opposites—working intensely and relaxing—the frustration of failure and elation of discovery—the disappointment at realizing you've been on the wrong track and the satisfaction of seeing all the pieces fit together. It requires imagination, intellect, intuition, and an aesthetic feeling about the rightness of things.

Solving a math problem necessitates overcoming a difficulty in an imaginative way. Sometimes this is done intuitively without conscious awareness of the creative process. Then you know you have the solution but can't explain how you got it. When you do know that you have solved a problem, you are directly in touch with the originality in your work.

Creativity can be seen in all aspects of solving math problems. It can even be seen in the different ways people do arithmetic and in the variety of schemes people invent to count on their fingers.

FIVE: THERE IS A BEST WAY TO DO A MATH PROBLEM.

A math problem may be solved by a variety of methods which express individuality and originality—but there is no best way. New and interesting techniques for doing all levels of mathematics, from arithmetic to calculus, have been discovered by students.

Teachers and textbooks may give the impression that they are offering the best way to do a particular type of problem. But two teachers rarely explain the same thing in the same way and the presentation in a textbook is just an expression of the author's way of doing mathematics. Even multiplication and division are done in different ways in different countries.

When several people work on the same problem, each may feel that the other's method is better:

Gloria: I did the comparative shopping problem by long division.

Barbara: Oh, you did it the right way. I just estimated it.

Gloria: Your way seems easier, you got the answer without bothering to divide.

Barbara: But I couldn't get the exact answer.

The way math is done is very individual and personal and the best method is the one with which you feel most comfortable.

SIX: IT'S ALWAYS IMPORTANT TO GET THE ANSWER EXACTLY RIGHT.

The ability to obtain approximate answers is often more important than getting exact answers. When a complex mathematical problem is "solved" on the computer, the answer that is obtained is approximate. The "exact" solution often is unknown and cannot be determined. But it need not be determined because the approximation is good enough for all practical purposes.

> Joan: After being married for ten years, I just found out that when my husband totals a check in a restaurant, he's not doing it penny by penny. He's sort of rounding it off and making sure that it is approximately seventeen dollars, or whatever.
>
> I thought you had to make sure the whole thing was exactly right. That made it much more intimidating. When I found out he was just sort of knocking it off, I got kind of mad because all these years I had been attributing a higher level of computational ability to his performance in restaurants.

Feelings about the importance of the exact answer often are a reversion to early school years when arithmetic was taught as a skill. There, the emphasis was entirely on the answer. This led to the feeling that you were "good" when you got the right answer and "bad" when you did not.

Just because a problem has an exact answer does not mean there is always great value in finding it. It depends on the circumstances. A waiter must give you the correct check, but you only need to see if it is "about" right, because it doesn't really matter if it is off by a few cents. You can even use your intuition. When a check is wrong, it tends to "feel" wrong and that feeling can be used as a clue that it is worth checking the addition.

SEVEN: IT'S BAD TO COUNT ON YOUR FINGERS.

There is nothing wrong with counting on fingers as an aid to doing arithmetic. In fact, most people find this to be very helpful.

Finger counting is often prohibited by parents and teachers and made fun of by peers. Those who use their fingers feel as if they are "cheating" or that they "shouldn't have to":

> Marge: If you count on your fingers in public, it shows that you can't do it in your head. Then everybody knows you are a dolt.

To avoid embarrassment, people do it surreptitiously—behind their backs, under the table, or by tapping on their legs or even on their nose—and feel guilty about it. The use of flash cards and speed drills for multiplication and addition reinforces the idea that finger counting is "bad" and that math should be done by rote.

Counting on fingers actually indicates an understanding of

arithmetic—more understanding than if everything were memorized. The abacus is really a sophisticated finger-counting machine that provides a fast and accurate aid to doing arithmetic. The Chinese use the abacus freely and feel no guilt about doing so.

Examination of many of the finger schemes people have invented for themselves has shown them to be clever, imaginative and efficient. There is no reason to prohibit their use.

EIGHT: MATHEMATICIANS DO PROBLEMS QUICKLY, IN THEIR HEADS.

Solving new problems or learning new material is always difficult and time consuming. The only problems mathematicians do quickly are those they have done before. But they are often expected to be able to add long columns of figures quickly, solve complex problems with hardly any thought, and have all formulas at their fingertips. Mathematicians don't have these abilities and don't expect that they should.

Time-limited tests, flash cards, and arithmetic drills have all added to the impression that mathematical competence and speed are the same thing.

> Annie: When I have to pause for a second to consider and don't know the answer just like that, I feel like I'm stupid. I get angry at myself for going through too many steps. There is a little voice that says to me, "God, you're slow!"

Speed is not a measure of ability. It is the result of experience and practice. A concert pianist makes the performance of a difficult program look easy because he or she has studied and practiced for years. But mastery of a new piece still requires work. Mathematicians do not expect to solve new problems quickly without the use of pencil and paper. The time it takes depends upon the number of similar problems they have solved, how confident they feel, how long it takes to get intuitive ideas —and a little luck.

When you expect to be able to solve a problem quickly and find that you can't, you tend to get discouraged and give up. There is no way to predict how long it will take to overcome the difficulty inherent in a problem. Once solved it may look "easy," but that does not mean that it was an easy problem or that it could have been done quickly.

NINE: MATH REQUIRES A GOOD MEMORY.

Doing math is like speaking a foreign language. When you are fluent you don't think about vocabulary or grammar. It is part of your thoughts and feelings. Knowing math means that concepts make sense to you and rules and formulas seem natural. This kind of knowledge cannot be gained through rote memorization.

The emphasis that is placed on memorization in doing multiplication and addition gives the impression that math requires a good memory. Later, it is natural to try to memorize procedures and formulas in algebra. But it is difficult to remember things you don't understand. As problems become more complex, memorization becomes increasingly difficult, if not impossible.

New concepts always take time to learn and it is easy to get frustrated and say to yourself, "I don't understand it and never will. I'd better memorize the rules." But not understanding a concept on the first attempt does not mean that a later attempt will fail. Everyone who has learned math has experienced a feeling of not being able to understand. It simply takes time and it really helps to try more than once.

TEN: MATH IS DONE BY WORKING INTENSELY UNTIL THE PROBLEM IS SOLVED.

Solving problems requires both resting and working intensely. Going away from a problem and later returning to it allows your mind time to assimilate ideas and develop new ones. Often, upon coming back to a problem a new insight is experienced which unlocks the solution.

It is generally felt that the way to learn is to make up your mind to do it, sit down, and then work hard. This may be true for adding a column of figures but it is not true for learning new concepts or solving math problems.

Resting is different from giving up. If you have worked intensely on a problem and think, "It's hopeless. I'll never get it," then you will leave the problem and completely stop thinking about it. If you think, "I can't get it *now,*" then you will continue to think about it both consciously and unconsciously. Mathematicians—like writers and artists—often experience flashes of illumination during the periods of rest that alternate with working intensely. Henri Poincaré (1854–1912), one of the foremost mathematicians of all time, wrote:

> For fifteen days I strove to prove [a theorem]. I was then very ignorant; every day I seated myself at my work table, stayed an hour or two, tried a great number of combinations and reached no results. One evening, contrary to my custom, I drank black coffee and could not sleep. [During the night] ideas rose in crowds; I felt them collide until pairs interlocked, so to speak, making a stable combination. By next morning I had established the [theorem], I had only to write out the results, which took but a few hours.

ELEVEN: SOME PEOPLE HAVE A "MATH MIND" AND SOME DON'T.

Do people really have different kinds of brains? The expectation is that those who have math minds grasp math quickly, easily and naturally. Concepts and ideas are mastered effortlessly. Problems are solved with barely a moment's hesitation. Correct answers just pop into their heads. If this is what you think, then when you look at a problem and don't know what to do immediately, you attribute it to being stupid or to not having a math mind. You get discouraged and say, "Why bother?" But the math mind is a myth.

Belief in myths about how math is done leads to a complete lack of self-confidence. But it is self-confidence that is one of the most important determining factors in mathematical performance.

We have yet to encounter anyone who could not attain his or her goals in math once the emotional blocks were removed.

TWELVE: THERE IS A MAGIC KEY TO DOING MATH.

There is no formula, rule, or general guideline which will suddenly unlock the mysteries of math. But it is helpful to see math for what it is and for what it is not. If there is a key to doing math, it is in overcoming anxiety about the subject and in using the same skills you use to do everything else.

CHAPTER THREE

what's at stake?

Math anxiety means that any contact with figures can produce an unpleasant emotional reaction and become a reminder of what you feel you can't do. It is natural to try to avoid math so as not to experience these feelings. If a person is claustrophobic and intensely fearful of elevators, the fear can be avoided if the person chooses not to live in a city with high rise buildings. Math and numbers are more difficult to avoid because they are everywhere. It is necessary to estimate costs, count change, balance a checkbook, check the bill in a restaurant, figure tips, get the most for your money by comparative shopping, and change the proportions in a recipe. It may be important to compute gas mileage, mark-ups, markdowns, discounts, sales taxes, and income taxes. The only way to avoid these things is to either not do them or have someone else do them for you. Even then, they remain constant reminders of things you can't do yourself and for which you must depend on others:

> Jenny: I have problems with bread. If I want to make one and a half loaves, I have to call my husband to find out how to change the recipe. . . . I used to work in a showroom. If a

buyer came in and I had to add up what they bought, which I fortunately almost never did, it would be a catastrophe.

Eileen: When the bank statement comes, I have this terrible sinking feeling. I expect to spend four or five days with it. Finally a friend said, "For god's sake get a calculator!" That helped, but I still had to press the right buttons.

Developing comfort and facility with figures makes daily tasks involving personal finances, recipes, percentages, and tipping simpler and more expedient. Newspapers and magazines take on new dimensions when graphs and charts are taken in rather than passed over. The business section of the newspaper may become more interesting. Before making a major purchase you may decide to compare technical data in consumer reports as well as prices. More important, you will feel increasingly independent and self-confident. You won't ask someone to help you because you are incapable, but rather because you are too busy.

The benefits of overcoming math anxiety go still further, because there is a widening of career options and improved job performance. The limiting effects of math anxiety on future jobs and careers come into play at the first opportunity to choose between taking and avoiding math. This usually occurs in high school. English is required for four years, while math usually becomes optional after algebra or geometry. Students who are not college bound may take business math instead of algebra. It is easy to see the importance of literacy, but the advantage of an expanded knowledge of mathematics is less clear.

On civil service tests that do not require algebra, it has been found that those who have studied algebra and geometry have a twenty-five percent advantage over those who have not. The reason is that taking more math increases "mathematical maturity," which is the ability to learn complex ideas and concepts with increasing ease. The more mathematically mature you are, the easier math seems. For example, mastery of algebra will result in a deeper understanding of arithmetic and how to solve arithmetic problems than could be achieved by studying arithmetic alone. "Mathematical maturity" is not inherited but is developed through study of more and more mathematics. For this reason, taking high school math courses results in improved performance in aptitude tests—even if the particular subjects studied seem irrelevant.

Many employers require prospective employees to take general tests that are half math—even when the available position requires little math. This is true whether the job requires a college degree or not. Lack of mathematical background or fear of math may cause you to turn your back on the job or do poorly on the test. This can be extremely disappointing if the opportunity itself is important to you. Phil clearly remembers having this kind of experience:

> I needed an average score of 120 on a battery of tests to get into naval air cadet school. My final score was 119. Every score but math was between 122 and 138. My math score was below second grade level. It was incredibly frustrating, I just sat there and cried. It was a tragedy to have missed that opportunity to fly by just one point.

The admission requirements of most colleges, graduate schools, business schools, law schools, and medical schools now involve taking standardized aptitude tests that include math.

Most people hate these tests, but if you suffer from math anxiety, you may feel that they are an insurmountable barrier:

Steve: I was applying to graduate programs and I was trying with all my heart to maneuver myself so I wouldn't have to face math. I took the graduate record exams and when I didn't score well I tried to finagle my way out of it. They just said, "We can't admit you, but if you raise your score we'll take you." I knew I would never get in.

Leslie: I wanted to go to business school. I bought a book to review for the business boards and did all the verbal parts. Every time I got to the math part I just had to close the book. I've got to get over this mental block if I'm ever going to take the business boards.

The frustration here is clear. It is infuriating to be stopped before you can begin to prove your capabilities in a field that interests you. As math anxiety is reduced, you will remember more than you expect and become more confident with the math you know. Concentration will improve and studying math and preparation for exams will become easier. Instead of becoming immobilized by panic, you will find that you can take tests with more comfort and utilize your full abilities.

Admission to college is possible despite poor scores on the math part of the scholastic aptitude tests, particularly if the verbal scores are high. It may not, however, be possible to get into the college you want.

Many people select a college on the basis of whether it has a math requirement or not. In college, math anxiety may lead to choosing a course of study based on avoiding math rather than real aspirations. Majors that have a math requirement may immediately be eliminated and a less interesting alternative may be chosen instead.

> Roz: I avoided anything that entailed math. I ended up getting my degree in psychology, which I really was not interested in, because it didn't have any math requirement at my school . . . I had wanted to get a degree in nutrition, but that required chemistry.

> Annie: When I started college, I thought I would be pre-med and take chemistry and physics and all that. One summer I thought I would go through a chemistry book to learn what basic math I needed. I opened it and never got past the first two or three pages. After that I majored in history . . .

Roz and Annie remained unhappy about these decisions that had been based on fear of math. After overcoming their math anxiety, they decided to go back to school; Roz to study nutrition and Annie to obtain a master's degree in business administration.

Math is becoming important in a widening circle of disciplines. Engineering, physics, and chemistry traditionally have required a considerable mathematical background. The expanded use of the computer, statistics, and mathematical models has resulted in math or statistics being required in many college programs in such fields as biology, economics, sociology, psychology, social work, and even political science, history, and journalism.

Businesses are finding computers increasingly useful and it is often necessary to read computer printouts and communicate with the computer division. But computers can not be used to full advantage if people fear asking questions. You can ask for the information you need even without knowing how to program or operate a computer. All that is required is a low level of math anxiety.

Most jobs are beginning to require some background in math or statistics. You may not be required to do calculations, but it is likely that the ability to read graphs or charts and make up budgets will be important. Statistical trends may have to be examined so that decisions can be based on future projections. Sometimes this can be overwhelming.

Ira: A few years ago, I worked on a salvage ship and had to signal how fast the deep-sea divers were to be brought up. I had to read charts and was so nervous that I thought I would misread something and kill someone. It was too much for me and I quit.

Jessica: I left a job in research because I was required to do financial analyses. If it made a big difference and it was the final thing on which they were depending, I just couldn't handle it emotionally. It got to be too much.

Anxiety about numbers means that working with them produces constant tension. This becomes much worse when others are depending on you. It is enough to make a person refuse a good paying job, resign from a position that is otherwise satisfying, or turn down an important promotion. Even when it is possible to get around doing calculations, there may be a constant fear of losing respect or being fired.

Joan: I can do fund accounting perfectly adequately when I am by myself. But if I have to do it with a member of the board of directors who is good at math, I just sit there going "uh huh! uh huh!" hoping that she is right. I check it after she leaves.

Sarah: If all of a sudden my boss calls down for some statistics, I freeze. I hand it to one of my assistants to do. One day I'm going to be found out.

When math anxiety has been overcome, both everyday tasks and work are easier. There is an increased willingness to work with figures, and calculations do not have to be checked and rechecked for fear of having made a mistake. At work, reports can be given and discussed with confidence and communication about quantitative data improves.

Many people have found that when they are comfortable with math, they become more assertive in asking questions. They find that others have the same questions and that people they always thought had all the answers don't necessarily have them. It may even become easier to question salaries and to ask for raises. There is a feeling of increased self-confidence that leads to seeking promotions to management positions.

Assessment of your real aspirations becomes possible when math ceases to be a barrier. When there are quantitative requirements, it becomes easier to fulfill them. And you gain real flexibility and freedom in your choice of goals.

PART TWO

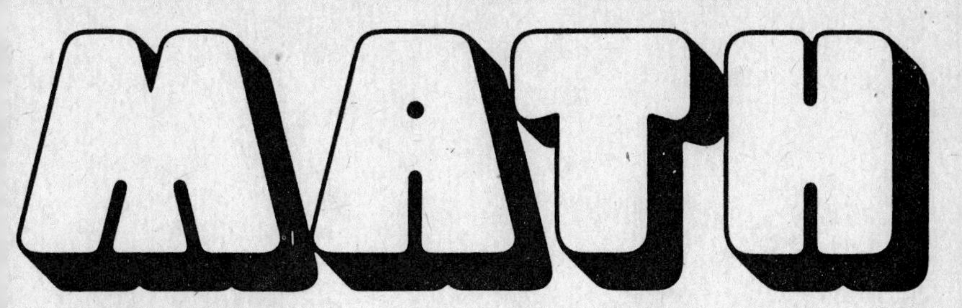

dynamics

CHAPTER FOUR

who's in control?

Knowledge of the nature and causes of math anxiety aids in accepting that it is an emotional, not an intellectual problem. If you are convinced that your difficulties with math stem from a lack of innate ability, you get discouraged before you start. After all, if it really were a deficiency then there would be little that could be done about it.

With the realization that math anxiety is an emotional problem that many share comes the feeling that there is hope. This problem can be overcome by first developing an understanding of how math anxiety affects you and then working through the various ways in which it determines your approach to math. Understanding alone is not enough. There must be a thorough grasp of the many facets of the interactions between your feelings and your attempts to do math.

Now that you have read Part I, you may have begun to think about some of your own experiences. Did you find you could identify with many of the comments people have made?

Thinking about the details of your experiences and feelings about math will lead to an increased level of awareness concerning how you approach math. Perhaps you made many important decisions in the past based largely on the desire to avoid math.

Many of the myths of math may have seemed terribly true and caused you to be unduly harsh on yourself when you found that you were having difficulty with math. Now you will become increasingly conscious of how often during the course of a day or week you come into contact with math or numbers. Begin to take note of how you react to math each time you are faced with it.

It is a good idea to discuss some of these things with close friends. You will discover that it is easy to find people who feel the same way you do. If you can't think of anyone you care to talk with, then writing down your own math autobiography would be very helpful. Write about how you felt about math in elementary school, junior high school, and high school. When did you do well in math and when did you do poorly? Do you remember any particularly traumatic experiences? Was there ever a time when you liked math? What changed that? Who helped you when you had difficulty with math in school? What did your parents think of math? What about your brothers or sisters?

The process of writing or discussing math feelings and experiences enables you to gain new insights into how you relate to math. You will remember things long forgotten. Thinking about things without either discussing them or writing them down leads to rumination—the same thoughts go through your mind over and over again and you feel like you are going around in circles. *Talking or writing breaks these cycles. It is the beginning of gaining control over your approach to math.*

Many different reactions to the suggestion that you discuss

or write down math feelings and experiences are likely. You may already have found people to talk to and feel supported by them. It may be that some people have seen you reading this book and have made discouraging comments like, "What are you doing that for anyway." Or you may be sitting and reading and taking all this in and not have had the chance to talk to anyone about it. You may even say to yourself, "Yes, of course, I will talk to someone." But you know that you never will. It may seem like a terrific idea to write a math autobiography, but you know that you will not get around to it. Perhaps writing is even more hateful than math.

The ways in which you relate to these suggestions are as important as following them. You may already be aware that you are not going to do them, even though you may think that the ideas sound good. Writing or talking about math feelings may seem like a useless exercise, because you feel that you have already thought things out in your head.

It is important to become aware of what it is that you will or won't do. If you don't like to write and know that you won't write anything down, it is better to admit this than to say it sounds like a good idea and never quite get to doing it. It is all part of discovering, accepting, and asserting your own learning and working style. This enables you to find ways to work that coincide with your own style and to make small changes that would amount to large improvements. It is another part of gaining control.

Toward the end of the first workshop session people sometimes say that they want to see some math. Some feel that being able to do math really involves knowing the "tricks" of math. It is easy to get the impression that math is nothing more than a collection of formulas and computational shortcuts. When someone shows you a terrific way of solving a particular type of problem it often does seem like a trick. And people often seem to take relish in showing you some clever way of doing something. They leave out how they came to their shortcut. It may have taken them a long time to learn or they may have come to it by accident.

Personally, we do not like these "tricks" or shortcuts. They may work but it is often difficult to understand the reasoning behind someone else's method. If you use a method which you don't understand, you will tend to feel uncomfortable and lack confidence in using it. The biggest trouble with shortcuts is that they rarely transfer from one type of problem to another. Each new problem seems to require a new method.

The quickest and easiest way to do a problem is the way that works best for you because you are confident and comfortable with it. We are concerned with general approaches to math. Ways of approaching math will be developed that apply to a wide variety of problems and carry over to seemingly unrelated things that have little to do with math.

If you feel the desire now to work with actual math, it may be an extension of the optimism that comes with finding that your feelings and experiences are understood and shared by many. Perhaps you really can get over this long-standing problem, now that you're ready to give both math and yourself another chance.

The most constructive way of using a mathematical problem is as a vehicle for gaining insight into your patterns of approaching math. It should not be seen as a way to learn tricks or as giving specific information on problem solving. Rather, the insights attained through looking at your own reactions to math will lead to an increasing ability to control your approach to math. Both talking and writing things down will help you gain control.

In the second Mind Over Math workshop, we hand out math material so that the group members can examine and discuss their reactions to it. This same material will be presented in a few pages.

Just thinking of math may cause you to get tense and apprehensive. You may feel that you want to leaf through the pages to see what is coming. It is better not to do this but to notice what you feel and think at the mention of math. You may already find your mind is beginning to wander. It may be getting harder to read this.

Can you imagine what your reaction would have been if you had seen math on the first page? You probably never would have bought the book.

When we first mention hand-outs, there is a visible rise in tension. There are jokes and uncomfortable laughter. Some people become very quiet and others change the subject. Some want to "just get on with it." Others would like to avoid it entirely.

It may seem like there is no purpose in doing all this talking before looking at anything. You may think, "I can't see how this is going to help me solve my problems with math." But once you look at math, it becomes more difficult to recapture exactly what anticipations you had.

Each time you know that you are going to have to deal with math, it is likely that you are going to have a variety of reactions. These reactions can completely control how you approach math if you don't fully understand and express them. You do not have to be in a math class or math anxiety workshop or reading a Mind Over Math book. You can be in a restaurant or a shop when you dread having to figure a tip or check a bill, or on the job when you know there are going to be some figures you will have to look over. Whatever the situation, there is likely to be an expectation of difficulty before you even start.

We are going to present a story about some things we did during a typical day that involved numbers. These include eating in a restaurant, buying gas, and purchasing supplies.

When it is time to hand out this story in the second workshop, there is an immediate change in atmosphere. The relaxed, comfortable living room begins to feel more like a classroom just before a test is given. Everyone moves around uncomfortably, talks in whispers, and helps himself or herself to a second cup of coffee.

Some people look through their pockets or pocketbooks for a pencil or pen. Others ask if they can use a calculator. We say it doesn't matter because we really don't expect anyone to solve any problems. The only purpose in handing out material is to give everybody a chance to notice and discuss their reactions to it.

Anticipating math can be a lot like anticipating a math test. The two often seem inseparable because math has long been associated with the pressures of performing and being evaluated or graded.

On math tests you usually have to do problems quickly and accurately. You may remember being called to the blackboard and having to stand there until you got the right answer. You may have been punished for getting the wrong answer. You may have had repeated experiences of being called upon to give answers when you didn't know them or were too nervous to figure them out. Memories of any or all of these experiences can be reawakened when you are faced with numbers or with a problem to solve. Why not jot down those memories that math brings to mind for you!

Writing down your feelings and thoughts can have a calming effect. You may want to do this in the margins or on a separate piece of paper. If you really don't like to write, you can just circle or underline those feelings that you think particularly apply to you.

When you feel like you are about to be tested you may start trying to imagine what the questions will be. You may think, "Uh, oh, here it comes," or feel that you would rather not have to look at anything. There is a tendency to expect the worst and think you're not going to be able to do anything at all. All kinds of thoughts can go flashing through your mind. How am I going to measure up? Is everyone but me going to be able to do it? Are they all going to think I am a dummy?

It is hard to concentrate when you expect the worst. Words and numbers seem fuzzy and can't be taken in objectively. Despite our advice earlier, you may find yourself rereading and wanting to skip ahead. This feeling is often expressed by saying, "Let's get it over with already."

You can't fail the material that is going to be presented because all you have to do is see your own reactions—all the thoughts, memories, and feelings you have related to math.

The purpose of this discussion has been to help you understand how you feel. Your expectations can be so negative that the first thing you see may bring on the feeling that you just can't do it at all—as you expected. Then you may give up. Negative anticipations stop you from ever really giving yourself a chance.

Overcoming math anxiety calls for experiencing and looking

at your emotional responses to math over and over again. You can work through an emotional block to doing math only by gaining both an emotional and intellectual understanding of how the block works and how it inhibits performance. Just thinking about it is not enough.

Some people adopt a wait-and-see attitude toward the presentation of math. They describe themselves as "just curious" and "only wondering" what the material is going to be like. They may even say they are looking forward to the "challenge" because they want to see if they can do it. It is important to realize that this response is also like preparing for a test. It is a test that is "challenging" and, if you fail to meet the challenge you set for yourself, you may start feeling discouraged.

A neutral response to math would involve little anticipation. This is especially true if you know you are not really going to be tested or evaluated. Your primary concern would probably be in listening carefully to instructions so you could fully take in what the real expectations were. You would then follow the instructions literally. If you began to have difficulties, you would start asking questions. In fact, even on most tests in school or college it is usually quite permissible to ask questions. The worst that can happen is that the instructor will say a particular question is not one that can be answered.

The more insecure you feel the more likely you are to blame yourself and think you should know and shouldn't ask. But when you are given a task to do, you have every right to ask clarifying questions so you can do your best without misunderstanding.

As we have said, the story we are going to present is about some of the things we did during a typical day that involved numbers. There are no questions asked. The idea is to read it and make up a problem of your own from any part of it. Be sure it is a problem you can solve. Most importantly, keep track of and write down all the feelings you have, even what you are thinking right now. Do you feel confused? Would you like to be able to ask us some questions? Are you having trouble reading this? Are you concerned about how you are going to do? Do you feel blocked? Do the instructions seem unclear?

Imagine yourself sitting in a room with us and ten other people who have been expressing feelings about math that are similar to yours. The couches and armchairs are arranged in a circle and there is nothing to remind you of a school. You have just been given a pen, a clipboard, and a sheet of scrap paper.

Take as much time as you like to look over this story. Don't hesitate to reread or take a short break between readings. There is no rush and no need to think anyone will look at what you are doing.

June 30, 1977

I really can't stand the heat today. It must be around 90° right now. Just a few days ago it was only 63°. I remember the heat wave we had in April; one day it was two degrees below freezing and the next it jumped to 95°.

I had some errands to run this morning before driving into Manhattan from Queens. We had run out of stamps, so I went to the Post Office and bought two rolls of 13¢ stamps, one hundred to a roll, and a dispenser for 5¢. I also went to the copy center to get ten copies made of a recent newspaper article. They were 7¢ for the first five copies and 6¢ for each additional copy.

I had to stop for gas and only needed 4.8 gallons which cost $2.85. I realized I had gotten pretty good mileage on my VW since I had gone 144 miles since my last fill-up.

Joe and I met for lunch at "La Garbage." Joe ordered a rare steak for $4.85 and coffee for 35¢. I had a chicken salad plate for $3.25 and coffee. The waiter put everything on one check, but we decided we would each pay for what we had ordered. We left about a 15% tip and went to a meeting.

Have you looked at the material for as long as you want? In a workshop, you would notice a long period of silence following the handing out of the story. You would start to wonder what other people were thinking, doing, and feeling at this point. You might even think about how long you should be taking and what you should be doing. There might be questions you would like to ask. You would feel relaxed or anxious, pleased or annoyed. It might have been difficult for you to focus and concentrate on the story itself.

After some time, maybe five or ten minutes, we ask if people would begin to share what they were thinking and feeling. Everyone participates and the discussion is very lively. People are very supportive of each other. What would you say if you were sitting in this group?

Annie: I still feel math shouldn't be part of the universe. Why is it here to bother me? After the material was handed out, it got quiet and my immediate feeling was that everybody knew what to do and was working, except me. I was just feeling discomfort and pronounced adrenalin.

Eileen: Every time I read it I got shakier and shakier. I felt worse than if it were the real thing, like a checkbook. I said to myself, "All I have to do is figure out how much each one spent that day." Then I thought, "That is a silly little problem, it can't be what they meant." But since that was all I could do, I decided I had better do it.

Roz: I felt a headache coming on. Then I started getting a little warm. I said to myself, "Gee, everyone is working." I couldn't figure out what it was that we were supposed to do. I don't think I even listened to the instructions.

Annie: It bothers me when I don't just know how much 3.25 and .35 comes to. When I have to pause for a second and consider, rather than knowing it just like that, I feel I am stupid. I feel I'm going through too many steps and get angry at myself. Then I just don't want to do it anymore.

Jessica: Why isn't anyone discussing the gas problem?

Arlene: I can't even look at it. There's all that stuff about mileage and gallons. And there are too many numbers there. What I do is read the words and skip over the numbers. I say, "Oh, great, they took a trip and they had lunch." When there is a number in the middle of a sentence my mind goes, "Delete, delete, delete!" I just go on to the next word. Then I start feeling a complete lack of motivation for it. I think, "Oh, well, I guess I can do something else with my life that has nothing to do with numbers." It's defeatism.

Sarah: If I had a calculator, I would keep punching the keys until I got something that looked reasonable. But it still wouldn't be a result of my reasoning.

Jessica: When I looked at it at first, that thing came down and I began to feel despair. Then I realized it really didn't matter if I did it right or not. I looked at it again and saw that everything was okay except the one with the gas. I knew I could have figured out how much it cost them to go that 144 miles. But I really would have to think about it for a very long time . . . like maybe a day!

Whenever I'm confronted with numbers, except with things that don't matter, it's like a sheet of white metal that comes down between my eyes and my brain and I cannot function. I just stop. Everything stops. I can't see the numbers at that point. I just blank it out.

Noreen: At first I got angry and thought the content was so dull. I said to myself, "Who cares about those chicken sandwiches." Then I noticed that everyone seemed to be doing something. I thought, "They're making up a problem What is the problem? The problem is this whole thing."

Then I read on and thought, "Well, so what!" and "La Garbage" summed up what I thought the whole project was. Then I got into analyzing the kid who eats the steak, what the poor chicken salad fellow felt, and the disparity between the price of chicken salad and steak.

Finally, I thought, "What are we supposed to do here? Oh yes, we're supposed to be making up a problem." I thought, "Well, there are lots of things in life you can make up and can't solve." So I said, "If you needed 4.8 gallons after driving 144 miles, how much did his tank hold originally?" No way could I solve that!

Jenny: I could never have thought of all the things Noreen did because I was literally suffering so much with the numbers. I just thought, "Oh God, I have to make up a problem." My head was feeling tight and I couldn't think. I probably could do it alone but I couldn't do it in the group. I couldn't believe I was suffering so much.

Eileen: Well, I'm just glad that nobody found another problem besides just adding up their expenses. I thought the trick was that there was another problem that I didn't see.

As these comments indicate, the reaction to the material can be almost anything. It can bring on discomfort or despair. It can make you shaky or give you a headache. It can make you angry or bored. It can cause real suffering.

If you are feeling anything like this, then it is no wonder you would want to avoid math. Who wouldn't want to avoid something that could evoke such painful feelings?

It may be that it is having numbers and words mixed together that causes anxiety. If you skip over the numbers as Annie does, then it is just about impossible to do anything with them. This is math anxiety at work.

Making up your own problem can be very disturbing. You begin to worry about whether you are doing the right thing and if you are making up the "right" kind of problem. When you are with a group of people, it is always distressing to see others working when you feel you haven't the slightest idea of what to do. You may prefer to have things more structured and to be told exactly what it is you have to do.

You may like the material even if you found some difficulty with it because you feel it applies to everyday life. It may be that math usually seems terribly useless to you and seems to be something you will never need.

When people find the material clear and don't have difficulty with it, they sometimes think it is a waste of time. It may seem senseless to work on something you already know. You feel the way to learn math is to be challenged.

You may see that there are decimals on the page and know

you don't like them. You may notice percentages and think about never having understood them. The fear may be that there are going to be things that you don't know.

Anticipatory anxiety is often so great that it is difficult to hear and fully take in the directions. Think about whether you remember the directions that preceded the story. After thinking about it for a few minutes, most people remember they were asked to make up a problem and write down their feelings. They may be uncertain about whether they were supposed to solve the problem they made up.

The part of the directions that is usually forgotten is that you were to make up a problem that would give you no difficulty; you were to be sure it was something you could solve.

Most people go to something they can't do. Rather than stay with an easy problem, they begin to focus on one that gives them difficulty—like the part of the story that has to do with getting gas for the car. Even when they start with something they find easy, they don't feel comfortable stopping, so they continue until something is found that gives them trouble. The things you can't do act like a magnet and draw your attention to them.

On math tests you will often get involved with a problem that seems beyond you. You keep working on it and working on it, feeling you can't move on until you get that particular problem. It really does have a magnetic effect on you.

Once you start focusing on what you can't do, anxiety builds up, takes control, and you begin to feel discouraged. You may even start thinking about all the difficulties you have had in the past. Then you give up completely.

Focusing on what you are afraid of almost guarantees those fears will be realized. It's like having to go to the front of the room in school. You begin to feel nervous and think, "I'm going to look awfully foolish if I walk into the waste basket." You try so hard to avoid it that you either walk into it or walk into the desk.

A basic principle of problem solving is to proceed in steps from easy problems to more and more complicated ones. You work from the obvious to the obscure. If you start with a problem you don't know how to solve, you will get more and more frustrated and discouraged. The solution, at first, is too far from your knowledge and confidence to be approached.

If you think about a big problem for too long you will get immobilized by the feeling that you can't do it. This applies to everything, not just math. When you first begin to play the piano, Mozart appears impossible. On your first ski weekend even the beginner's slope looks like it is vertical.

When you proceed gradually, step by step, you gain more and more control. You build on small successes until you achieve your goal.

Mathematicians always start with an easy problem, one that they can solve. The harder the problem they have to work on, the more they gravitate toward an easy one. For example, if either of us were given a research problem to solve, at first we would have no idea where to begin. We would then change the problem.

The idea would be to rephrase or simplify the problem to the point where it was manageable. One way of doing this would be to make up a related but easier problem that we would be sure we could solve. After solving such a problem we would reexamine our work. The purpose in doing this would be to gain confidence and insights into the larger problem.

Suppose, for example, you wanted to figure out how much

gas cost per gallon but were unsure about whether to divide or multiply. In fact, you really didn't like having to work with numbers like 4.8 and $2.85. Ask yourself, "What would make the problem easier for me?"

Could you do the problem if it had been five gallons for three dollars? Then the cost would have been sixty cents per gallon. But if you know how to do this problem, then you can also approach the problem you didn't like. The question you could ask yourself is, "How did I get sixty cents?" Think about this for a few minutes or do some work on scrap paper. After a while you would realize that you must have divided 5 into $3.00. That means you should do exactly the same with the numbers you didn't like, divide 4.8 into $2.85.

If you found you did not follow this explanation, then you should reread it slowly and get some scrap paper. Trust yourself and give yourself the time you need to think it through.

Another objection may be that you never thought of doing it that way and don't think you could have come up with it yourself. But the important point is to realize the necessity of gaining control through starting with what you do know.

One other concern may be that you have difficulty dividing decimals. If you know that is where the main difficulty lies, you also know that once you learn it you will be able to do this kind of problem.

When we first started writing material for the workshops, we wanted to do something that would bring a guaranteed, successful experience to everyone. Our reasoning was, if we told everyone to make up a problem they could do, they would feel very relieved and each person would find something in the story they were comfortable with. This was not the case and the dialogue that was just presented is typical of what happens when people look at our story.

A literal interpretation of the instruction to make up a problem you could solve might have led you to ask a question such as, "How much did the temperature rise from a few days ago to today?" or "How much did Joe spend for lunch, excluding tax or tip?" You might then have felt satisfied with having followed the directions in a straightforward way.

Doing math requires confidence and concentration. Each word, number, and symbol must be clearly taken in and analyzed. Panic and anxiety make this impossible and can easily control your approach to math. As you start to recognize your patterns of reacting to math, you will find it gets easier to take a second look at problems. And you will discover that you see more than on the first reading. You will find your concentration improving, and you will be amazed by what you can do when you give math your full attention.

CHAPTER FIVE

math games we play on ourselves

How often is your mind really calm? Most of the time we have little conversations going on in our heads. Internal voices tell us how to feel (don't be nervous, stop worrying), and how to behave (be serious, stop being silly, control yourself). They compliment us (that was great, you did a good job, you deserve to celebrate) and criticize (don't be so clumsy, that was dumb, that was stupid). We tell ourselves when to slow down (stop rushing, take it easy, there is plenty of time) and when to speed up (you're too slow, get a move on, don't take all day). We even urge ourselves on (stop procrastinating, get to work) and give ourselves warnings (be careful, you're going to make a mistake, don't forget).

Internal dialogues are often helpful, but sometimes it is impossible to calm yourself down or get yourself to work right away. While pep talks may be useful, being hard on yourself is not. Attacking yourself and calling yourself names just makes you feel bad and immobilizes you.

When faced with math, the internal dialogues become both negative and defeatist. Thinking things like "I don't have a math mind" or "Everyone knows what to do, except me" lead nowhere. They are self-defeating games—games you play on yourself. If you know what these games are, you will be able to catch yourself playing them. This will help you to stop.

"EVERYBODY KNOWS WHAT TO DO, EXCEPT ME."

A horrible feeling associated with math comes from the experience of taking math tests when you panic and are unable to think. You find you can't even get started. As you look around, it seems that everyone else is working, and you think, "Everybody knows what to do, except me." They all seem to be working very intently and started writing as soon as they got the test in their hands.

The question you should ask yourself is, "What are they working on?" Often, the first thing people do when they get a test is simply to recopy the questions. It helps calm them, but they are not actually solving anything. When Annie told the group that she had felt everyone else was working and knew what to do, they all laughed because they were having the same thoughts. Sometimes people start writing because they can't bear the feeling of sitting and doing nothing. What they are writing may have nothing to do with the problem. If you are having difficulty, you can be sure you are not the only one.

"I DON'T DO MATH FAST ENOUGH."

It is upsetting that it takes so much time to do math. Should you be able to do addition by knowing what the sums are immediately? Very few people can do that! But if you can't do a problem quickly, just like that, without thinking, you may feel, "I don't do math fast enough."

Speed and skill at math should not be too closely linked. Speed is a separate skill that is developed through repetition and practice. Learning new concepts and doing new problems is always a slow process.

Everyone has their own pace of doing things. Some talk fast, others slowly. Some walk very rapidly, others very leisurely. One way is no better than another; they are just different styles.

Changing your normal pace is very difficult. Fast talkers find it very hard to talk slowly. Slow talkers stumble over their words if they are forced to talk rapidly. If you like to mull things over in a careful reflective way, you will find speed pressures particularly burdensome. Too much time pressure can result in a marked decrease in performance.

Fast addition, for example, comes through practice. That is why the grocer who writes the prices down on the brown paper bag can add like a flash. Think about how many of those problems he does in a day. Suppose you were to do two hundred addition problems daily, every day, for years. Don't you think you would be able to add quickly then?

You have to ask yourself what you really have to do fast and what can be done at your own pace. College boards and graduate record exams do require speed, but it really does not matter if it takes you a few extra minutes to check the bill in a restaurant.

If speed is of real importance, then it can be attained by first gaining understanding and then practicing. Almost paradoxically, the best way to gain speed is to slow down. This is particularly frustrating if you are used to working quickly. It is like learning to type. You can't teach yourself to type fast. First you have to learn the basic system and then you gradually increase your speed by typing a lot. If you try to work too fast, you will make too many mistakes. This is especially true in math. Not only will you make errors but you will start feeling very confused. It is better to do problems at a comfortable speed and do them correctly than it is to rush and make mistakes.

Some things are learned very quickly and others seem to take forever. Usually, it is impossible to tell what makes the difference. Eventually you can learn all the math you need. How long it takes is not important—and it will take much less time than you think.

"I'M SURE I LEARNED IT, BUT I CAN'T REMEMBER WHAT TO DO."

Suppose you are faced with a problem that you can't solve. If you once knew how to do it, you begin to get frustrated. Often you experience a rush of associations and all sorts of rules and formulas pop into your head at once. But you have no idea which one is correct. For example, if you have to work with fractions, which you know you always hated, words like *common denominator* and *invert* might come to mind but you don't know what to do when.

There is no reason to expect to remember how to solve problems you haven't done for years. Suppose you had studied French for years and become quite good at it. If you didn't speak or read it for three or four years you would expect to forget a lot. The next time you tried to speak you would find it very difficult. It would take review and practice before it began to come back. The same is true with math. Like anything else, you forget it when you don't use it and it begins to come back if you review it again.

"I KNEW I COULDN'T DO MATH."

If you see a percentage or fraction you may think, "I never understood fractions or percentages." Rather than taking a positive approach and saying, "I can do everything but fractions and percentages", you say, "I knew I couldn't do math." You give up when you could say, "It is something that I will have to look up and learn."

When reading the *New York Times* you always come across unfamiliar words. If it is not possible to understand the meaning from context, you say, "I will have to look that word up." You could skip the sentence entirely and look it up later. Even if you never got around to it, you wouldn't think you didn't know how to read and decide that reading the *Times* is impossible.

If you have to write a paper or an article, it is necessary to have basic references like a dictionary, thesaurus, and grammar text at hand. The same is true if you have to do any math. It is extremely helpful to have books that review arithmetic and algebra handy. Such books are readily available. It is easy to forget math. As with any subject, there are always some things that you don't know. If there is a word, symbol, or computation that is unfamiliar, you should not hesitate to look it up. Why not treat math as you would anything else?

"I DON'T HAVE A MATH MIND."

If you can't do a math problem right away, it can feel as if there is something wrong with you and you think, "I don't have a math mind." When you feel this way, that leads you to think, "I would rather not have to look at it." You feel that you are not going to understand and expect to feel dumb. One way not to feel dumb is not to look at the material. This is a motivation for avoiding math. If you are coaxed into giving it a try and don't know what to do *immediately,* you feel confirmed in your belief. You say, "I knew this would happen." Your belief confirmed, you give up. If you tried again, you would discover that you could do more than you expected.

When you can't get a problem, your head starts feeling cloudy and thinking gets muddled. You even begin to feel sleepy. Your brain feels as if it has stopped working. This happens to almost everyone. If you see it as a deficiency in yourself, then you are sure to stop trying.

Finally, you ask someone to show you how to do it. You just want to be shown what solution to memorize. When questions are asked in this way, the answers are never really taken in. You can't follow what is being said or remember the answer and the feeling that you can't do it is reaffirmed.

The best way to learn is from your own difficulties and mistakes. But if you are self-critical, a correction feels like another criticism. This makes learning from your mistakes impossible.

Those who do well in math expect to feel muddled when they start and don't get discouraged. They don't concern themselves with how long it takes them to learn a concept or solve a problem. You should not expect to be able to do math quickly.

And even if you can't do something, you still can learn from your mistakes.

"I GOT THE RIGHT ANSWER BUT I DID IT THE WRONG WAY."

Even when you get the right answer to a math problem, you may be unhappy with your work. Sometimes you feel you were just lucky and could not do it again. You think your method is too complicated. Someone else's method must be better because it seems faster or more clever in some way. This means you don't give yourself credit for what *you* have done. Instead of taking satisfaction from having gotten the right answer, you take credit away from yourself by criticizing your own work.

There is no best way to do a math problem. In fact, it is rare for two people to do the same problem in the same way. When teaching math we have often done the following: After presenting the solution to a homework problem, we ask if anyone has done it by any other method. Invariably, someone will offer a somewhat different approach that leads to the same result. In fact, in a class of fifteen people there will usually be fifteen variations in the solution to any problem.

There is no value judgment placed on different solutions. Just the opposite; the creative part of doing math is in the variety of ways people have of doing problems.

Rather than looking for the originality in your own work, you look for its faults. You say, "It was too roundabout" or "I used arithmetic when I should have used algebra" or "I counted on my fingers." None of these things matters. What is important is that you have found a way of working that is successful for you.

"THIS MAY BE A STUPID QUESTION BUT . . ."

Most questions that are asked about math are prefaced by the phrase above. Often you will not ask a question because you feel it is something you should know. This feeling is reinforced by the confident style in which math is usually presented. It seems simple but you know you do not understand. You feel inhibited and think you are the only one who has such a question. The assumption is that if someone else had that question, it would have been asked.

As teachers, we always assume that when one person asks a question, he or she is asking for many people. If you don't ask for fear of appearing stupid, then you can't gain the understanding you need. Worse still, you feel once again that there is something wrong with your mathematical abilities.

When someone asks one of those supposedly "stupid" questions in a math class, the usual feeling on the part of the class is relief. Most people have the same question and the same reluctance to ask.

"IT'S TOO SIMPLE."

Math is expected to be hard. If a problem seems easy, you become suspicious and think you have missed something.

Often people look at a problem, realize they know how to do

it, and then don't bother because it is easy. The same is true when reading a math book. These books always start by discussing simple examples and then build up to harder ones. It is through a thorough understanding of the easy problems that you gain insight into the harder ones. Solving problems you are able to do can make you feel like you know something. By saying they are too easy, you take this away from yourself.

When looking over a math book, do you read the introduction? Do you start from page one as you would with other books? Many people turn to the end of a math book before they look at the beginning. The material there is bound to appear difficult and cause them to say, "This is too hard." Then they turn to the beginning and say, "This is too easy." If it is something you *can* do, it is too easy; if it is something you *can't* do, then it is too hard. In either case you come away feeling bad. If you understand the first few pages of a math book, that means you have the background for it and it is the right book for you.

Even when you find you understand something you may think, "This is okay but what I really don't understand is . . ." or "It's okay for now but I know it's going to get complicated." As soon as one question is answered you skip to another thing. It is better to take satisfaction in knowing you have learned something you didn't understand, than to rush on to something else you don't know.

"MATH IS UNRELATED TO MY LIFE."

When you begin to have difficulty with a problem, you question why you have to go through it. You think, "Who cares about this anyway?" These kinds of feelings lead you to set your goals so that math really is not a part of your life. It is possible to do this by carefully choosing a career that doesn't involve math. But you may be eliminating work you would really enjoy. When it comes to daily things, like balancing your checkbook, you get someone else to do them for you. Why lose free choice of careers and become dependent on others?

CHAPTER SIX

math games others play on us

There is a great deal of outside reinforcement for the math games you play on yourself. Others are playing math games on you, both intentionally and unintentionally, when they say things such as, "That's easy" or "You will never be able to do math." Such comments are wrong and cause you to feel criticized and discouraged. In no way are they helpful. It is as important to recognize these games as it is to recognize the games you play on yourself. When you do this you will find that they lose their negative effect on you.

"YOU DID IT THE WRONG WAY."

This game is the counterpart of one you play on yourself when you undervalue your own method. The other person hardly even looks at your work before dismissing it. They then say they will show you a "better" way to do it. It is a myth that there is a best way to do a problem.

If your method works for you, then it is not the wrong way. Even when you make a mistake, you are usually giving the right answer to the wrong question. For example, a seven-year-old girl was asked to add 12 and 23. The result she got was 8. The teacher marked her wrong. But how wrong was she? What question was she answering? She added one and two and two and three to get eight. She did this correctly. She just didn't answer the right question.

It is most helpful when someone enables you to find out what question you are answering. Just being told you are wrong only serves to make you feel bad. If you learn what questions you are answering, you are also able to learn the right questions to answer.

"YOU SHOULD KNOW THAT."

If you ask a question, this is the one thing you don't want to hear. This attitude can be conveyed nonverbally through tone of voice or facial expression. Questions like, "You mean you really don't know that?" make you feel like you should never have asked. Teachers sometimes say, "You should have learned that in the ninth grade," or "That's high school math." But if you knew it and had learned it you wouldn't have asked.

You have a right to ask when you don't understand something, whether you should have learned it previously or not. Instead of commenting on your question, it would be far better for someone to either answer you or suggest a good reference where you can learn it yourself.

It is important to think about whom you ask for help. If someone constantly gives you the feeling you should know the answer to your own questions, then it is better to ask someone else. The message that you don't know what you should is very discouraging and prevents learning. Whether this message is given intentionally or unintentionally is unimportant. You should seek out the people who will help you without making value judgments.

"YOU WILL NEVER BE ABLE TO DO MATH."

Many people recall being told this. It may be said in a critical way or as a simple statement of fact. You are told to accept that math is a subject you just can't do. There are two opposite reactions to hearing that you don't have the ability to do something. Some respond with an attitude of "I'm going to show you!" and do everything in their power to prove they can do it. The more common reaction is to stop trying, especially when a parent or teacher is the one who tells you.

Saying that a person doesn't have the ability to do something has the effect of self-fulfilling prophecy. You then interpret difficulties as confirmations of your supposed lack of ability. It causes you to give up rather than try again. When you have doubts about yourself, failures confirm those doubts. It becomes hard to accept that it is only a temporary problem.

Some mathematicians report having always done well in math. Others can recall having had difficulty in both high school and college. We know many mathematicians who got C's, D's, and F's in math courses in college. They did not begin to do well until they got to graduate school. If you are having difficulty at one point, it does not mean you will always continue to have the same difficulty.

After overcoming their anxiety, many people in our workshops have gone on to study math they had never been able to do. Some have taken algebra—the same algebra they absolutely "could not do"—and have had no trouble learning it on their own. Others have taken accounting or calculus and received A's.

Math instructors sometimes make the unintentional mistake of telling their class that the course has a reputation for difficulty and that twenty-five percent of the students normally fail. They may then say that if everyone works hard there is no reason for anyone to fail. This may be the instructor's way of trying to inspire everyone to pass by working hard. But if you start off being afraid you can't do the work and are told that many fail, you are likely to give up and say, "If anyone is going to fail it is going to be me."

Being told something is going to be difficult can be an incentive to work if you start off feeling confident of your abilities. It leads to total discouragement if you start off doubting yourself. We advise teachers to avoid making this kind of statement.

"IT'S OBVIOUS."

Math books and teachers make liberal use of phrases such as "It's obvious" or "It easily follows." Is it really obvious or is it something that is hard to explain? Math research articles are famous for saying "It follows," when to follow it might take two or three pages of work.

These phrases are very misleading. Often, only after thinking long and hard about something will it begin to seem obvious. Try this experiment with a four-year-old child: Line up ten pennies in a row, count them, and then rearrange the same pennies into a pile. Ask the child to tell you which is more, the pennies in a row or piled up. He or she will say the row is more because it is longer. It may be obvious to you that they are the same, but it is not obvious to a four-year-old.

There is a story about a famous mathematician who was teaching an advanced course to graduate students. He said that something was obvious and was questioned by a student. He looked again at what he had done, excused himself, worked feverishly in his office for fifteen minutes, came back and said, "I was right. It is obvious."

A problem always seems very clear and the solution "obvious" *after* you have solved it. Supposedly obvious things are the most difficult to explain.

"THAT'S AN EASY PROBLEM."

If you've been struggling with a problem for a long time, it is very upsetting to be told this. People seem to relish letting you know how easy they found something. The hidden message is, "I could do it easily, what's the matter with you that you can't?" This reinforces the feeling that you can't do math.

People rarely tell you how much difficulty they really had with a problem. They don't say whether they worked with someone, how long it took them, or how they got the idea that enabled them to solve the problem. They don't reveal whether they had seen the problem before and don't talk about how many similar problems they had done.

Instead of being told that something is easy, it would be better if you were helped to locate the point at which you were stuck and then guided past your difficulty.

If you have had a lot of difficulty on a math problem before you solved it, you will find that other people have had trouble at exactly the same points. You have located the trouble spot of the problem. After some time, it gets increasingly hard to understand why you couldn't do it right away. You forget what that "trouble spot" was.

The people you think are talented in math have more difficulty than you realize. Sometimes when you ask people questions, you find they don't know nearly as much as you expected and are not nearly so comfortable as you imagined. They may even have had trouble with math courses themselves!

"ALL YOU HAVE TO DO TO LEARN MATH IS TO WORK HARD."

We are forever being told that success requires a lot of hard work. Since math is seen as a logical subject, it is usually assumed that emotions play no role in doing math. The solution to a problem may not have an emotional component, but emotions affect the work all along the way.

People who are math anxious often recall working extremely hard at math without being able to get the right answers. But most of the time may have been spent trying to calm down enough to concentrate. These efforts lead to intense frustration because anxiety must be overcome before you can work effectively.

The feelings math evokes cannot be ignored. They must always be recognized and dealt with. When frustrated on a problem you can acknowledge that you feel like giving up. But you don't have to give up. You need to take a break and be prepared to start over again.

CHAPTER SEVEN

realistic expectations

Wouldn't it be nice if math anxiety suddenly disappeared? It is what we all hope for. Whenever faced with serious problems, we want to have some magic key that will make the difficulties go away. We hope for a wonderful insight that will unlock the doors of mystery. Perhaps, you think, learning some magic tricks will enable you to solve problems.

This is not the way it happens. The change is always slower and less dramatic. Small changes in the way you approach math begin to occur. Perhaps you ask a question when you never asked one before. Maybe you try to figure out some small thing, like a restaurant check that you never did before. You begin to notice that the people you always thought were so terrific at math really are not. You may start to balance your checkbook or look at numbers more carefully.

Every small change is a big one. If you have been doing things all your life in a way that has not been working, then any change is significant. These changes are often played down as not important. That is wrong. Not only is it wrong, but it works like a math game you play on yourself, because it is self-defeating. You do not give yourself credit for what you do. Rather, what you don't do comes in loud and clear.

Sometimes you ask for the explanation to some math question. When it is answered, or just at the point when the answer is beginning to get clear, another question pops into your mind. In this way you move toward what you don't know, rather than toward what you do know. Thinking about what you don't know is bound to make you uncomfortable and anxious.

Another way of undoing success is to keep focusing on all you have to do. This is done when you start thinking about all the math you need to learn. It seems so overwhelming. What you might like to do is to erase all the past math and start all over again with a new subject, like calculus, without bothering with all the old stuff. Would you think of learning French III by skipping French I and French II?

When there is a lot to learn it seems like it will take forever. This overwhelming feeling also makes it hard to assess your needs realistically. For example, if you realize you have not mastered arithmetic skills involving division or fractions or decimals, then math feels like a hopeless task. But as an adult it becomes easier rather than harder for you to learn these things because of your years of practical experience. It becomes easier to relate what you are learning to your life. The material covered in the first eight years of school can be comfortably contained in a single, not too fat book. It takes months, not years to review what you have missed or forgotten.

When a writer must produce a three-hundred-page book and has written only fifty pages, the feeling of how much there is left to be done is overwhelming. He can easily become so intimidated that he can no longer work. He tries to imagine what is going to go into another two hundred and fifty pages. That is impossible to do all at once, so the task seems insurmountable. It is very difficult to imagine how the words are going to come that are going to fill up so much space.

The only nonformidable task is writing one page. That seems much more manageable. So keeping things going really requires looking at one page at a time. Any larger images can halt the whole process.

This applies to math. If you have ten problems to solve, think only about doing the first. Once you get this you can move on to the second and the third.

Don't be surprised if you notice that when you are confronted with math, you are reminded of a math class, a math teacher, or a math test.

It is easy to get discouraged if you notice you are still doing some of the same things and reacting in some of the same ways to math. But regressions are normal. There is nothing wrong with slipping back. The only thing that is required is that you take note when this is happening. This will provide the extra push you need to get back on the track.

There is no reason to think your feelings toward math should have been neutralized at this point. Math anxiety is overcome a little bit at a time. No one can just hand you the secrets. You need to be a participant and engage in the process.

It is important to be realistic both in the goals you set for yourself and in assessing what you can do. For example, if you know there is a particular time of the day when it is difficult for you to work, it is better to say, "I'm not going to be able to get any work done," than to kid yourself into thinking that you can do the impossible. You will surely be disappointed with yourself, which will make it even more difficult for you to work in the future. On the other hand, if you say, "I know I am going to be able to work" during such and such a time, then you will be able to constructively plan when it is that you are going to work.

If you are a writer and know that you can write say, five pages a day, then it is foolish to set a goal of ten for yourself. You are sure not to meet it. Then you won't have any satisfaction from doing six pages, which is one page more than you usually do. The same is true for math. If you know it takes you two hours to study ten pages, there is no sense trying to do twenty pages in that time.

It is necessary to be realistic in the kind of challenges you set for yourself. Many people believe the way to learn is to be challenged. If you feel this, you must be realistic in the challenges you set for yourself. They must be just at the edge of what you know. If they go too far afield, you get so overwhelmed that you get discouraged and give up.

Once anxiety is overcome, maturity is a plus in learning. Everyone needs to learn things they have never learned or have forgotten. People who come to a new field at any age often make major contributions based on a wealth of experience that enables them to come up with fresh approaches.

It is unrealistic to expect to work with something you don't know. You can't work with percentages if you don't understand them. But knowing what you need to learn allows you either to look it up or to ask someone how to do it. It doesn't matter whether it takes you one hour or ten hours to learn. After you have learned, you never think about how long the learning took.

PART THREE

CHAPTER EIGHT

sherlock holmes

Now is a good time to look at some more material and see what kind of response you have.

The title of this chapter may have started you guessing what it was about. It is a Sherlock Holmes story and, as usual, Holmes has figured something out. The answer is given. You should see if you can figure out how Holmes solved the problem.

What do you think of Sherlock Holmes stories? If you like them you will probably be quite calm and looking forward to this one. If you hate mysteries in general and Sherlock Holmes in particular, you may already be dreading what is to come and find yourself tensing up. These anticipatory responses are perfectly natural. Remember, the main purpose in looking at this story is to gain insight into your own responses and learn more about yourself. This is more important than being able to solve the problem.

You will find that you are more in touch with what is happening to you than you were with the first presentation of a math problem in Chapter 4. As before, the most important thing is to notice and write down your reactions. This makes you an active participant in the process of learning to do math. The more active you are, the more you will discover you can do.

Give yourself lots of time to look over the story, write down your feelings, and think about your responses. Reread it as much as you need to. This is not a test of any kind, so time is not a factor. And of course no one will look at your work.

In London there were three gangs operating on August 11, 1891. Holmes knew from some inside information that his equal in crime, the clever Moriarty, led a gang with five members. At the same time the treacherous Smerzi headed a gang with seven members and Gilda Z., the trickiest of them all, a gang with eight members.

Watson: Have you figured out whose gang pulled off the Great Train Robbery of October 3, 1891?

Holmes: Yes.

Watson: But how, Holmes?!!

Holmes: Elementary arithmetic, my dear Watson.

Watson: Let me in on how you did it.

Holmes: From certain information from Scotland Yard, it was known that originally none of the gangs was large enough to pull off the Great Train Robbery. They must have added another organization that was twice the size of the original gang. Altogether, twenty-one members were involved in the robbery.

Watson: This is all too much for me. I hate math. I can't do it. I block and get anxious. Just tell me whose gang did it, I can't figure it out.

Holmes: The treacherous Smerzi's gang.

Although you may be tempted to read on at this point, try not to. Even if you feel as if you can't look at this story again, give yourself another chance. Think about what you do and don't understand about it. It is important not to give up too soon.

The Sherlock Holmes story is given out during the third Mind Over Math workshop. There is plenty of time to read it and think about it and most people help themselves to a second cup of coffee before they begin to do any actual paper and pencil work. After about ten minutes we begin discussing the responses people have been having. What was your initial response to this story? Did you feel like any of these people?

Sheila: There are certain things I always skip reading. Sherlock Holmes is one of them. I first thought, "Oh, God, I'll never be able to do it." I skipped over it the first time I read it. Then I very consciously forced myself to read it again.

Steve: My reaction was just the opposite. I avoided looking at the material last time but this was like a game. I love Sherlock Holmes.

Annie: I got a headache. I felt angry that there was no gang with 10.5 people. My reaction was, "Okay, I've spent my two minutes on it. Can't do it. Finished. Let's have the next problem."

Loretta: When I first read it I saw "train" and started thinking about those train problems where two trains are going in opposite directions. I said, "Oh, no!" Then I felt good because I took my time, didn't give up, and came out with an answer.

Peggy: I felt tricked by the dates. I felt you wouldn't have put them there unless there was some significance. I wasn't as fast as I would have liked.

Harry: I was confused about the gang leader, thinking that maybe he counted and maybe he didn't.

Julia: It seems to me that in math there is always some trick involved. All the men in my family are very good at math and like to play math games. I didn't know how to do the Sherlock Holmes story and felt much stupider than last time. I fiddled with the figures and nothing came out. It felt like one of those games they play in my family.

There is no intentional trick in the way the problem is written. But it is easy to feel tricked when you know there is something you are not getting.

You may be getting impatient for the solution to the problem. In the next section, we will discuss a way of approaching it which will both calm you and help you avoid getting caught up in the wording. This approach will have application to all word problems. But first, it is essential to continue to explore the emotional process of approaching problem solving.

Sometimes even when you make good progress with a problem there is a tendency either to lose confidence in yourself or to fail to give yourself the credit you deserve.

Eileen: I felt very anxious but I think I did it.

Jenny: I felt more relaxed than last time. I knew no one was going to take my paper. And if I made a mistake, it wouldn't be too bad. But I still couldn't get it. What I want to know is how Eileen figured it out.

Eileen: Well, I read a lot of murder mysteries so I liked the problem. I tried all sorts of calculations. It was not until the fourth time around that I realized what it was.

Sarah: First I added up all of the members including the gang leaders. Then I scratched that and added up the gangs and got 20. I said that doesn't work.

Noreen: First I decided that it was absolutely impossible. Then for some reason I wrote down A is for answer and A plus two A is 21. Then I wasn't really sure if that was really the answer. It was the first equation I ever made up.

I figured out that there were three of those A's altogether. If I divided 21 by 3, I came up with 7, which is the number of one of the gangs. I was not sure what I was doing, but something popped.

Ruth: I did the same thing, but then I got worried. I started thinking about whether Smerzi was a member of the gang.

Sheila: I just thought, 7 is a number and doubled it to get 14. These numbers totaled 21. I didn't even think about why and certainly can't explain it.

What was your reaction to the Sherlock Holmes story? Were you able to solve it immediately or after several tries? Did you do it the way Noreen did or the way Sheila did or another way entirely?

Remember that there is no best way to do a math problem. Anything that works is fine.

Perhaps parts of the story seemed unclear or tricky and caused you to give up. It is easy to feel intimidated by those who have solved it and to feel you could never do it yourself. That is not true.

Were you more relaxed or more tense than with the first material? Did you lose confidence in your own work? Are you satisfied with what you have done?

HOW TO READ WORD PROBLEMS

Word problems are confusing because there is so much that needs to be sorted out. When you read word problems with the expectation that you are going to be able to understand or solve them immediately, you are likely to be disappointed. They are not meant to be read that way. Math prose can seldom be understood the first time through.

Prepare yourself to read by doing something to relax like getting a cup of coffee. Then read as passively as you can. Read word by word with an eye toward deciding what is or is not important. Avoid anticipating what the problem is or how you are going to solve it. Don't think about math or about formulas. Just concentrate on one word or phrase at a time. Write down what seems relevant and tentatively discard what seems irrelevant. The problem and method of solution will often emerge without your focusing on it.

Look again at the Sherlock Holmes problem and consider a different style of reading it.

First you read, "In London . . ." Don't read any further but stop and say to yourself, "I really don't care where they are talking about." If you disagree with this and feel the name of the city may have some importance, then write it down on a piece of scrap paper. Circling or underlining works too, but jotting the "important" things down on a separate piece of paper works much better. It brings them out in sharp focus because it separates them from the rest of the material.

The next phrase is, ". . . there were three gangs . . ." This should be written down as "3 gangs." You may be tempted to try to remember that there were three gangs, but it is better not to rely on your memory. If you do, you will find that you have to keep looking back over the problem to refresh your memory. This can easily make you nervous because you start worrying about forgetting. There is also no reason to "clutter" your mind with things that can be written down. Leave yourself free to concentrate on the material itself.

At this point your scrap paper may look like this:

London 3 gangs

But if you thought "London" was unimportant, that would not appear.

Now continue reading, ". . . operating on August 11, 1891." You might guess that this detail is also not important and decide to ignore it. Or you can disagree and wonder how on earth anyone can know in advance whether the date is important or not. The fact is that all of these decisions have an arbitrariness to them. The best guideline is, "When in doubt, write it down." If some of the things you write down later turn out to be unimportant, you will just pay no attention to them. On the other hand, if you have left out something that you need later, you can always go back and write it down.

This type of reading is very slow, careful, and disciplined. You will find you experience the impulse to get on with it and rush ahead. But in the long run, reading in this way will become a time saver. It avoids anticipation of what the problem is going to be and requires that you go one step at a time.

You should avoid relating too strongly to the material in the story. For example, you may have decided in advance that since you can never figure out Sherlock Holmes mysteries, you are not going to be able to do this one either. Or it is possible that you went to London once and didn't like it, and you will begin to think about all the things that happened to you there.

The story should be approached with complete neutrality. Your only goal is to gather facts that may come in handy later. Don't try to figure out why you will need them or how they will fit together.

Now we read, "Holmes knew from some inside information. . . ." This is a warning that something important is coming. But there is nothing to write down. We are just breaking down sentences into fragments or phrases. Continuing, we read, ". . . that his equal in crime . . ." This is just a descriptive statement—not relevant to the math. Some people are bothered by the word "equal." Since it is supposed to be a math problem, there is the anticipation that "equal" is going to have some special meaning. It is also possible to start thinking about how Holmes can be equal to Moriarty since Holmes is not a criminal. This wondering about meaning indicates you are getting too involved in the story and are anticipating what is to come. Try to catch yourself doing this. Then remind yourself that the purpose is to gather information, not to figure anything out.

The story itself has to be kept at some emotional distance so you can focus attention on just writing down what you think is important. Don't think about the story too much.

Again reading, ". . . the clever Moriarty, led a gang with five members," you might think to yourself, "I don't particularly care how clever Moriarty was but I do want to keep track of how large his gang was." This can be done by abbreviating and writing M–5.

When you read about Smerzi and Gilda, write down S–7 and G–8 to remind you of how large their gangs were.

Now you have pulled out all the information in the first paragraph. The most you would have written would be something like:

London   3 gangs   August 11, 1891   M–5; S–7; G–8.

If you had felt the city and date were not important, you would have even less written down. The nice part of doing this on a separate piece of paper is that you can now take in everything that was in the first paragraph at a glance, rather than seeing a jumble of words that are hard to take in and work with. It is likely that you will never have to look at the first paragraph again. You have also given yourself something to be satisfied and comfortable with, because you have begun to understand and decipher the story.

Now comes the dialogue between Holmes and Watson: "Have you figured out whose gang pulled off the Great Train Robbery of October 3, 1891?" One way of responding to this is to think, "This is just a lot of talk," and continue reading. But if you had decided earlier that the date was important, you would also want to write down this new date.

What we are saying is, organize the mathematical information. Take out the numbers and put them down in an orderly fashion, dates with dates, gang numbers with gang numbers.

If there are facts of questionable importance, write them down too. If they turn out to be irrelevant, they will hardly be noticed by the time you get to the end. In this problem, you will see that the dates turn out to be of no significance. If you had written them down, you would stop noticing them by the time you got to the end of the problem.

There is more dialogue between Holmes and Watson, but it is just talk with no information worth writing down until Holmes says, "From certain information . . ." This is another warning that there is something important to come. Slow yourself down and read very carefully and deliberately: ". . . it was known that originally none of the gangs was large enough to pull off the Great Train Robbery." There is still nothing to write down, but it is important to think for a moment or two about what this statement is saying. It may be worth rereading it several times. Repeat to yourself, "originally none of the gangs was large enough."

It is tempting at this point to try to anticipate what is going to be said or asked. That is not helpful. It is still necessary to remain the passive reader who just takes down information. There should be no thinking about solving a problem until everything is carefully read and recorded.

Sometimes people take the numbers from the first paragraph and start manipulating them, trying to get them to arrange themselves in some way so that they come to 21. This number was noticed on reading through the problem in its entirety.

It is not wrong to try to manipulate the numbers, it is just too early.

Don't rush through the next sentence in the story but carefully read, "They must have added another organization . . ." Think about the meaning of this. Ask yourself who "they" refers to. It is saying that in order to pull off the robbery, one of the original organizations must have added others to it.

Reading on we find, ". . . another organization that was twice the size of the original gang." Perhaps you don't know what to do with that and want to reread it a few times. In fact, you don't have to worry about what to do with this information. You just need to write down the relevant numbers. These numbers represent twice the size of the originals. You can figure out what all the doubles are and write them down next to or under the originals. Again, you may not know how you are going to use this information, but you want to record everything. Your scrap paper should now have on it:

M–5	S–7	G–8
10	14	16

Then it says, "Altogether, twenty-one members were involved in the robbery," and sitting right there you see a 7 and a 14 which together give you 21. So you put a circle around S–7 14 and conclude that it must have been Smerzi's gang.

The approach we have taken to this problem has been first to collect all information, including whatever we have doubts about. It has been written in such a way that, at a glance, we can look at it and take it in. Once you have done this, you should give yourself a few moments to look at your information and absorb it. Think about it for a while. Do some scrap work on the side. The solution may not come immediately, but it will eventually if you don't allow yourself to give up.

Does the problem seem clear now or are there still places where you feel confused? When the Sherlock Holmes problem is discussed, there is always controversy over parts of the story and its interpretation. These controversies will be discussed next. Every problem is capable of generating a myriad of approaches and questions.

If you understand the problem now, you may think, "Why didn't I see it before? It's so obvious!" But math problems always seem obvious after they have been solved. That does not mean that it was easy or that there is anything wrong with you if you did not do it. Peggy Fleming makes skating look effortless. Does that mean it is really easy?

CONTROVERSIES ABOUT SHERLOCK HOLMES

Many people wonder whether the leaders were supposed to be included in the gangs. Here it is a question of wording and interpretation. This uncertainty can be handled by trying the problem with and without the assumption that the gang leaders are included in the gangs. For example, if you were wondering about this, as a trial you might have written:

M–6	S–8	G–9
12	16	18

Now 6, 8, and 9 represent the numbers in the gang plus one more for the leader. The numbers 12, 16, and 18 represent twice the size of each gang. Since none of these totals 21, you would reject the idea that the leaders weren't included in the original numbers, 5, 7, and 8. You would then try the problem assuming the leaders were included.

Very often you can resolve doubts about the way a problem is meant to be interpreted by trying it more than one way. Also, if possible, it is a good idea to ask questions if a problem is unclear to you. More often than not, unclear wording is inadvertent rather than intentional.

Many people get very involved in the story itself and focus on what is happening rather than on gathering and using the information the story contains. A new story may be made up in which the total comes out to 21 but all the given conditions are not satisfied.

> Rosemary: Smerzi's gang had 7 members two months before the robbery, excluding him. His treachery caused him to take over the other two organizations by knocking off the leaders, Moriarty and Gilda Z. His own 7 gangsters plus their 5 and 8 totaled 20. When you add Smerzi you get 21.

This is creative, imaginative, and gives the correct total. However, it does not meet the requirement that an organization twice the size of the original be added. A clue that you are getting too involved in the story is when you start filling in events that are not written down. If you catch yourself doing that, take a short break and try the problem again.

Adding another outside organization was felt by many to be very unfair.

Julia: I just felt that was a real dirty trick because you've only discussed three gangs and they were all the gangs that existed. But they don't fit the figures. So you tricked me.

Most people have experienced brain teasers or other math problems where there was an intentional deception. Nobody likes the feeling of being deceived and tricked. It is infuriating. When a problem is hard to figure out, it feels "tricky." But this reflects the difficulty of the problem and is usually not the result of purposeful deception.

You may sidestep the difficulty in conceptualizing an outside organization by focusing on the words, "twice the size of the original." That does not allow for the added gang to have been one of the originals. When you write down the doubles, 10, 14, and 16, you are forced to conclude that it couldn't have been one of the original gangs that was added because none of them was the right size. This is another advantage of jotting down all the information in the story.

When you expect to be tricked by math problems, you tend to mistrust the solution you get if it comes easily. This can cause you to give up when you have actually solved the problem quite successfully.

Steve: I read the problem and immediately wrote $7 + 14 = 21$. Then I thought, "It must be more difficult than this." So I decided to try to use algebra. Then I couldn't do it.

There is no reason why you have to use algebra to solve this problem. For this and all problems, whatever method you use is okay.

Sarah: I made up an equation and put $x + 2x = 21$. I even went so far as to put down $x = 7$ and circled the 7. Then I looked at the 7 and said, "Okay, now what?" I started to look back and figure out where they got the people from. I didn't even realize that I had the answer. It just made no sense at all.

Sarah couldn't believe she had actually done the problem and she had to look at it again. This just got her more and more confused. Actually, both Sarah and Steve had used their "mathematical intuition" to come up with a solution to the problem. This is similar to the intuition you use in daily life and is a fundamental part of solving problems. It is the subject of the next chapter.

self-awareness in doing math

> Fran: When I was working on the Sherlock Holmes problem, I automatically, without even thinking, wrote $7 + 14 = 21$. There was no thought process. I just wrote it down. I disregarded it and said, "Well, now I have to figure out how to do the problem."

Fran had solved the problem almost instinctively. Many people find that either $x + 2x = 21$ or $7 + 14 = 21$ "pops" into their heads. There is no conscious reasoning process attached to coming up with these thoughts. These intuitive math ideas are most helpful in problem solving. They should never be disregarded or discarded. Often the first thing that comes to mind will, in the end, turn out to be correct.

When teaching math and giving math tests, we often request that students write everything in ink. This is because there is a tendency to lose confidence in your first ideas and to erase them. It is just those things that need to be preserved. Students will often solve a problem correctly and then cross out the solution and replace it by an incorrect one.

When this happens, we cross out the incorrect solution and give credit for the crossed-out one. The purpose in doing this is to give credit for the use of mathematical intuition. These are the math ideas that first come to you. Everybody has this intuition and it needs to be encouraged and developed. You will have a "feeling" about the answer or method of solution to a problem long before you can fill in the steps.

> Beth: When I see columns of numbers, I do them fast and well but I have never understood exactly what I do. I have the feeling that it has to do with visualizing the shape of the numbers. For example, when I see three, I visualize three points. I'm an artist.

To further encourage the use of intuition, we recommend that math teachers not ask students to explain how they got an answer when the answer is right. In a math class, we may ask if anyone knows the answer to a problem. When someone offers the correct answer we then ask if anyone else can explain it. The person who gave the answer may offer an explanation if he or she wishes, but it is not necessary to do so. When you know the answer but don't know how to justify it, being asked for an explanation robs you of the pleasure of getting the answer right. These answers exemplify mathematical intuition at work.

WHAT IS INTUITION?

How many times have you said to yourself, "I thought that but never said it." Did you ever feel, "I knew from the beginning that I shouldn't get involved with him (her)?" Have you ever thought, "I knew I wasn't going to be happy on this job"?

Intuitive insights are based on the sum of your knowledge

and experience. The gut feelings you have provide guideposts that should always be taken into account. You feel things for a reason and these feelings will often point you in the right direction, in math as well as in life.

Self-awareness in doing math means allowing yourself to be intuitive. Being in touch with your feelings puts you in harmony with one of your greatest strengths. You use only half of yourself when you rely only on reason. The same is true when you rely only on feelings. Both always need to be considered. Whether you tend to be a more logical or a more intuitive person, don't forget that we all have the capacity to use both faculties.

Living only by logic and reason can cause you to neglect powerful emotional forces that are ultimately most important. Decisions based purely on reason often turn out to be wrong. What you feel is often more on target. But relying on feelings alone also has disadvantages. Sometimes your feelings are based on a misreading of people and situations. Some logical thought can easily bring clarity where confusion reigned.

UNCONSCIOUS WORK

Math, like any other creative human endeavor, is done partly unconsciously. Repeatedly looking at a problem or concept, going away from it, and returning to it allows your mind the time it needs to assimilate the ideas it is exposed to. Some call this "putting it on the back burner." It permits the development of new ideas.

Doing math is not a continuous process where you keep working and working. There is a place where you get stuck and start to struggle to get past a difficulty. Then you let go of it for a while, before struggling with it some more. At some point, all of a sudden, you know what to do. In drawing, your hand makes the right move. In writing, you suddenly know how to rephrase a paragraph so that the meaning shows through.

Something happens unconsciously that enables you to go the last step. When you just keep working on a problem without taking a break, you get into trouble because you are not allowing your unconscious to work.

Stan recalls:

> I went to another University to work for a year on research with a senior professor. He gave me a problem I just couldn't do. I couldn't do anything. Zero. Every morning I would go to my office, open a book, and give up. And I did my best to avoid running into the professor I was working for. It would have been awful if he saw me and asked me what I had done. I would have had to say, "Nothing!"
>
> Each day, for over a month, I would go to work, write one line, get stuck, and find someone to have coffee with. Finally, I decided I just had to face up to things and admit defeat. I called to make an appointment to see him. I was going to tell him I had not done anything.
>
> Just before it was time to see him, I decided the least I could do was to make a list of all the things I had tried and failed at. I wrote down the first thing and started to write down the the second when I suddenly knew how to solve the problem. I spent half an hour writing up my ideas and went to the meeting. I showed him my work and he said, "Great, this was exactly what I was hoping for!"

It almost seems magical, but it is not. All the efforts put in paid off in a most surprising way. The problem could not have been done just by sitting and working. It had to be done through a process that involved starting and stopping and starting and stopping. One day the solution was there.

If you work intensely on a problem, your mind will continue to work on it unconsciously when you stop working. Many people even experience waking in the middle of the night with the solution to a problem they were working on but were unable to unravel. This will happen as long as you don't take an attitude of totally giving up. As long as you say, "I will keep trying," rather than "I can't do it," your mind will work on it for you. This is not to say that all you have to do is read a problem and then take a nap and you will wake up with the solution. But periods of intense work, alternated with rest, are very productive.

Flashes of illumination are very exciting. You can visualize the process as first having to absorb large quantities of information. Then you allow it to be felt and conceptualized. You are never really walking away from it. You carry it around in your head. Unconscious processes continue to go on selecting the relevant from irrelevant. The process continues because the material is in the machine. Then the solution suddenly bursts into consciousness.

Barbara: I've done the same thing when I've written magazine articles. I'll just flounder around for weeks and the material keeps skittering through my mind. But I can't assemble it in any way. I feel like a fool. Then suddenly something somebody says makes the whole thing replay in my mind and take on a structure. It has happened to me with *everything I have ever written!*

Karen: It happens with other things. Sometimes I think about calling people and trying to set up appointments. Some days it is just not happening. I push at it and push at it and make more calls. Then there are more things that are not working out. I think, "Let me get away from this. I am not making another call today." I'll wait even another couple of days until I can try it and it feels right.

Joan: I never tried that with math because I always had the feeling that with math "close" is no good. "Right" is the only thing you are going for. Unless you are on the "right" track everything else is useless.

The answer is only the last step in doing math. The process of doing math is far more important and interesting than the answer itself. Making some wrong turns will help you to get on the "right" track.

RECOGNIZING INTUITION

The more mathematics you learn, the more sophisticated your intuition becomes. But you can be intuitive at all levels of mathematics.

Did you ever get the feeling you overtipped? Did you ever sense you were being overcharged in a store? Something in back of your mind says, "Think about it!" You walk out and say, "Something is wrong. Something doesn't feel right." It is a clue that you should double-check. More often than not, when it feels wrong it is wrong.

Karen: I bought a lot of things; pants, tee shirts, scarves. I paid my bill with a charge and didn't look at it until I got home. It surprised me how much it was. I checked the addition but didn't find any mistake. But the feeling was so strong that I just had to press on. I checked the tickets and found that one had been transcribed incorrectly by $15.

Ordinarily, I wouldn't have even looked at the bill. I had the absolute pleasure of really trusting myself and looking further after feeling defeated on the first try. When I went back to the shop they were very apologetic.

Some people know almost to the penny what the tape is going to say in the grocery. It is not something they consciously calculate. Try to see if you can develop this ability. What you do is look at your groceries before they are added up. Don't try to do any calculating, just guess the result. Notice whether you are too high or too low. The next time you go shopping do the same thing. Again, take note of whether you are too high or too low. Also notice if you are off by as much as last time. Don't criticize your own performance. Be a passive observer of your estimates. If you do this each time you go shopping, you will find there is a steady improvement in the accuracy of your guesses. And it is all done without any conscious calculation.

A similar ability to "guess" answers is true for algebra problems. Even if you don't know any methods, you are likely to be able to answer simple problems without doing any work. For example, if A stands for answer, what is A if $A + 3 = 7$? Can you "see" that the answer is 4? It is possible to offer a sequence of problems of increasing complexity in which you can "see" the answers. Learning algebra then means developing procedures that will work for very complicated problems. These methods are based on what you have already been doing very quickly in your head.

HOW TO USE INTUITION

The intuitive and the logical parts of doing math are the result of separate processes. Just as in writing, it is difficult to come up with creative ideas and grammatically correct organized structure at the same time. If you want to create something, you must first allow yourself to write down whatever you are thinking. Afterward you rewrite, reorganize, and correct grammar and spelling. When you try to do both at the same time you are likely to find yourself blocked and unable to do anything. Exactly the same thing happens with math.

> Joan: I find that when I am writing and I'm blocked, it works for me to just write down anything. I don't worry about sentences or structure or presentation or whatever. I just have to get something down and then I can start the process of thought. Later I can go back and rewrite it.

Mathematicians make liberal use of scrap paper. They are forever scribbling numbers, letters, and even pictures to help them visualize. They jot down all their ideas as they come to mind. Whatever is not written down tends to get lost.

The first step in solving a math problem is to allow yourself to mark down anything you think about the problem, no matter how irrelevant it seems. The ideas that come to you in this way turn out to be most useful. It is amazing how frequently your first ideas are right.

The first random words, numbers, symbols, and pictures that you think of represent the intuitive ideas you are having. Write them down and worry about corrections and refinements later. Remember, you can always throw away your scrap paper. Everything you write down is not necessarily going to be in the final work, but it provides a rich source of ideas which give rise to other ideas which then flow more easily.

Don't worry about how material is going to fit together. You can't see the end when you are only at the beginning. You can't make ideas come in the sequence in which you will finally present them. Your only concern should be to make sure you get all of them down. Logical presentation comes last.

> Phil: A funny thing happened to me last week. I was doing a percentage problem that I had to do. Somehow, I don't know why, I did it with fractions. I haven't used fractions in ten or fifteen years. My hand just did it. I didn't even have to think about it.

Trying various methods allows your intuition to work for you. If you get a feeling about something, try it. If it doesn't work, say "too bad" and try again.

At times your mind goes racing so fast that it is difficult to get everything you are thinking written down. When these thoughts are not recorded they are extremely difficult to retrieve later. What you can do is to write down a word, phrase, or symbol that you can refer to later. This will serve to bring back the memory of what you were thinking at the time.

Elaine: I wondered what method I was using. I thought, "Okay smarty, you came up with 7, but how did you get it?" Then I said, "I don't know. 7 times 2. That's not it. That's not the way. What's the method? There is no method." That happens very often. I don't know how I got there.

Don't make value judgments on your own thoughts and ideas. You will be able to separate the relevant from the irrelevant later. If you try to do that too early you will stop the flow of ideas. You stop yourself when you think, "It's no good. It's not the right method. Why on earth did I ever think of that!" You will find work more productive and more enjoyable if you can just be yourself without making judgments.

THE PLACE FOR LOGIC

The arithmetic and algebra you learn in school took thousands of years to develop. As we pointed out earlier, the logical, step-by-step presentation in math books is a condensation of the work of all those years. The logic may intimidate you because it seems to be all there. But there are many ideas that still don't "feel" right. That is because the intuition on which so much of math is based has been left out.

There is a place for logic, but it comes after the intuition. When you solve a math problem there is an immediate sense of relief. You look up the answer in the back of the book and say, "Great, I got it!" Then you want to get away from the problem as fast as possible. But there is another step after you get the answer. It is to do the problem over again and see if you can figure out exactly what you did to get the answer. This gives you a greater depth of understanding and provides the tools that can be used for other problems.

> Bob: I never stop to think about what I am doing. It was so hard to just try and do it. I would have a whole scratch paper full of figures and calculations. Once I had the answer, I never thought of thinking about what I did. My feeling was "Wow, I finally did it. Let's move on to the next one."

> Phil: I would just go through a problem. If I got it right, I never went back to think about how I arrived at it. I either knew it by rote and that was that or I never thought about it.

The final step in problem solving involves reorganizing, reordering, and selecting the best ideas. This consolidates what is known, in much the same way that a second or third draft of a paper markedly improves clarity and language.

Solutions have often been obtained through a series of intuitive steps. You can learn a lot about a problem by taking the trouble to do it again, this time paying careful attention to each step. You then try to find the logic that justifies your intuition. Ask yourself questions such as "What was I thinking at the time I did this?" and "What reason is there for doing it?"

The process of rewriting and providing logical steps is not fast. But the benefit is an improved grasp of the problem and of how everything fits together. This makes you more skillful at solving other problems. And you will be less intimidated by the neat ordered look of math books when you discover the many refinements you are able to make in your own work.

decreasing anxiety

Not all anxiety is debilitating. In fact, there is nothing wrong with being a little tense when you are doing math.

Have you ever noticed that you are most effective and creative in the presence of a little pressure? There is an optimal level of anxiety. Some pressure gets you moving. You focus and think more clearly and work more intensively. When you are totally relaxed, your mind often wanders. Too much anxiety can make work impossible, but a little can spur you on.

EXPRESSING ANGER

There are many things about math and your experiences with math that can cause you to get angry. When this anger is unexpressed and stays inside you, it has no place to go and turns into anxiety. This makes it impossible to work.

As you become more aware of your anger toward math, it becomes easier to express it. That helps decrease anxiety and also enables you to channel your anger constructively into increased assertiveness.

Wendy: I feel mad at the system that didn't give me the guidance and reinforcement that I needed as a child. I now realize I can do math. I resent all those painful hours in class that I never got anything out of.

I regret that nobody ever encouraged me to try to do well in math. Nobody seemed to care. I feel deprived.

Steve: Last night, I did a little bit in the GRE review book. I didn't spend much time, maybe an hour. I was able to do some of the simple things but the complex problems were really difficult. I got angry at myself for not being able to do them.

Too many value judgments have been placed on doing math. It is often presented as the ultimate test of intelligence. If you accept this, then you conclude that the reason you can't do math has to do with what is wrong with you.

The myths that have been perpetuated about math also make you feel you are doing something wrong when you count on your fingers, don't know how you got an answer, use your intuition, or can't do a problem quickly. So you get angry at yourself rather than at those who have perpetuated the myths.

Donna: It annoyed me that I wanted to understand it, and could have, but the author prevented me from doing so.

When you can't understand a concept, you get frustrated and feel helpless. Then you get annoyed at yourself, at the teacher, or at the book. But sometimes it is just that certain ideas are difficult to grasp. Blaming yourself and others does not lead anywhere. It is productive to be aware of your feelings and to give vent to them, but then you have to go past that stage.

Teachers are sometimes poor and books are often badly written. But that is true of all subjects, not just math. Try to remember good and successful experiences as clearly as you remember bad ones. There are many good teachers and friends who can give you the kind of supportive help you need. Try to find them and avoid those who are not helpful.

It is possible to see instructions in math as commands. Your math book doesn't say, "Would you mind adding up this column of figures," but only, "Add! Subtract! Multiply! Solve!" These feel like imperatives telling you what to do. No one likes to be ordered around. And you are told only that your answer is right or wrong.

Math does not have to be seen this way. There is an emotional satisfaction in doing problems in original, imaginative ways.

You will feel much better and work more productively when you channel your energies into increased assertiveness in asking questions and in not allowing people to play any math games on you.

Sarah: Last night I had a battle with my husband. We were discussing a testing program I had been working on. He started talking about how you could measure intelligence by testing the ability to do math. He said that whether you could do math or not was a simple question of ability. I snapped at him and told him I simply refused to believe that.

Nancy: I had to open a new account in the bank and had to ask questions about it. I did not want to slump down in my seat and be afraid because he was going to give me a percent. I made him explain everything to me.

I made a decision about what I wanted to do. I figured that if I made a mistake I could change it. I didn't feel so overwhelmed.

FACING THE ANXIETY

You cannot decrease anxiety by fighting it and trying to talk yourself out of it. It is better to accept that you are anxious. Allow yourself to feel the symptoms and fears. Do you usually experience anxiety in the stomach, chest, back, or head? Anxiety escalates when you start worrying about being anxious.

Is the anxiety you feel the same old math anxiety and not caused by other personal problems that are unrelated to math? Remember all that you have learned about math anxiety and how anxiety affects your approach to math. Begin to focus on getting yourself started.

What usually relaxes you? Do you like to have coffee or a soft drink? Do you like to listen to music while you work? Do whatever makes you feel best.

STARTING

It is easy to get into a vicious cycle in which you reinforce your worst fears and doubts. If you don't get started doing math, you get to thinking that you cannot. When you doubt yourself, the task becomes increasingly overwhelming.

Breaking this cycle requires that you begin anywhere. Don't try to find the ideal place to start.

Nancy: I like to paint. But sometimes I approach a blank canvas and just don't know what I am going to do with it. Then I start working and get an idea and then another and another.

The easiest place to start is with the first thought that comes to mind. This can be an intuitive math idea, an attempt at a problem that you feel competent about, or it can be a statement about how you feel toward math. If you have no ideas, try writing, "I have no ideas." It may lead to something else. You may realize that you do have some ideas. Once you get going, ideas start to flow.

STAYING WITH IT

When you look ahead to anticipate where you are going, it gets too confusing. You lose satisfaction because you are jumping to what you can't do. Stay with what you can do. That will be encouraging rather than discouraging.

There are times when you have to accept that you are having a bad day. You cannot expect to be equally productive at all times. Athletes have slumps, artists can't create, writers get blocked. Sometimes personal problems, not related to math, interfere with your ability to think clearly. Those days may just have to be written off. Accepting this makes it easier for you to work on the good days.

MEMORY

Have you ever forgotten the names of close friends when you have to introduce them to each other? It is most embarrassing. A few minutes later, when you are less tense, all the names come back to you.

When you forget because of tension or anxiety, there is no way to force your memory to work for you. You just get into a vicious cycle where you chastise yourself for not remembering. This makes you even more tense and less likely to remember.

> Noreen: Under normal circumstances I know the multiplication tables like a snap. Under stress situations, 8 times 7 does not equal 56. Once that starts happening and 8 times 7 is 42, I'm shook.
>
> I know that under times of stress I can deviate from things I know as well as my name. Then I start to kick myself. I question the basic precepts. I wonder who made up numbers and whether anything is really valid.

Until you relax a little, you are stuck. Engaging in some activity will help you calm down enough to get your memory going again.

You can respond to a situation in which you lose your memory with acceptance rather than self-criticism. Say to yourself, "I must really be nervous if I can't even remember what 8 times 7 is." Then turn your attention to working out the answer by whatever means you can. Once you are working, it will be harder for you to think about how nervous you are.

In times of tension often two different methods come to

mind and you can't decide which one to use. Your first idea is usually the more intuitive one. It is more likely to be correct.

When you are comfortable with basic math concepts you can figure out what you can't remember. It is easy to forget facts but difficult to forget concepts. For example, you can often figure out the values of the multiplication table more easily than you can remember them.

freeing yourself from the past

The first step in freeing yourself from all your past attitudes toward math is to come to grips with your feelings and reactions when confronted with it. As you gain awareness of what math does to you and of what you do to yourself, you will find it is not just math that sets off certain negative responses in you. With that discovery, you will become acutely sensitive to your own patterns of reacting and will be able to catch yourself and begin to change.

Sheila: I notice I often have the same reaction to other things that I have to math. I block things off and stop listening. I actually feel myself clicking off and not hearing. I realize I do this with anything that presents a challenge I feel I cannot handle. This is something new that I never realized I did.

When I first looked at the page that had percentages on it, I stopped when it said 15%. Whatever I read after that made no sense to me. When I realized I had tuned out, I also realized I could do something else. I reread it again and I actually understood it. It was the first time I ever reread a math thing.

Carol: I realize it is not only math, it is me. I have a very low frustration tolerance. In learning to ski I want to go up to the top and come right down like all the others. I don't want to start out slowly. I just want to be able to do it.

Phil: If it's not fast, I want to skip it. I want everything to be fast.

Peter: When I was taking several subjects at the same time I always did the easy things first. I would leave the math to last. Then I would say I didn't do the math because I didn't have the time.

You are freeing yourself from a past when anxiety was high and you got anxious about being anxious. At the first signs of difficulty you gave up. As you begin to take a more realistic view and notice changes in yourself, you start feeling less helpless, more in control, and more willing to keep on trying.

Sarah: I had to read a study and decided I was not going to panic. I was not going to do what I usually do and just push it off. So I read through it and then asked someone to explain the how and why of the things I didn't understand. I think the explanation went through this time.

Annie: Last week I had to get the prices for some furniture from the manufacturer. But then I had to compute a 40 percent discount on everything. I did some computations and showed it to the man who had asked me to do it. He said, "Okay, let's see."

He pulled out his calculator and got a different figure. There was an enormous difference. I got 94 and he got 150.

> When he said a different number than mine, it was like a shot of adrenaline going through me. I felt completely that he was right. Then I thought, "Is it time for me to think I don't know how to do this?" I decided, "No, I really did it right." I told him I would like to see *him* do it again! He did and found that my answer was correct.

REMEMBERING THE PAST

Try to remember when your fear of math began. Are your present fears rational? Math anxiety is like a phobia; the panic you experience is based on fears that have been with you for a long time. You can free yourself from their influence—but not by pushing them away. See if you can trace them to their roots. Were these fears rational in the first place? Would you behave differently if you didn't have them? How have they influenced the thoughts you have always had about yourself?

You need to remember these things and then think about the differences in you that exist in the present. You don't have to do things the same way you did before. You are not trapped by bad experiences if you learn from them. You can learn and reinforce new patterns of acting and reacting. Your past is part of you and cannot be undone, but it does not have to control you anymore.

One way of getting at what may be behind your math anxiety is to try looking at a math book. If you find yourself getting very tense and your concentration starts to go, write down everything you feel as it comes to mind. This may remind you of other situations in which you felt the same way. Doing this may enable you to free yourself to do math.

Ruby: Last weekend I decided I had better sit down and start reviewing the statistics I had a year ago. I had to review the first semester because I have to take the second. It is a requirement of the degree I am working on.

I decided to try writing my feelings down. I sat down with my stat book and this blank piece of paper and wrote down feelings as I went along. It went something like, "My head feels very tight. My mind is wandering. I think I'm getting a headache. My back is starting to hurt."

All of a sudden, the whole thing seemed too hard to do at all. I felt absolutely overwhelmed by the whole thing and started to cry. I thought, "This is really weird. I took the course, made a B+ and am crying because I can't understand the first three pages."

There was something wrong with my being so upset. I stopped, got a hold of myself, and said, "Wow! This is crazy."

All that overwhelming business reminded me of the last time I really had a bad time with math. Math erupted for me at the same time that my mother got very ill. I was in the ninth grade and she had to go to the hospital. The way I felt about math was really the way I felt about that situation. We never spoke about it in my house. We weren't a house to deal with feelings. I kind of zapped all the feelings I had about that situation onto math because the last time I remember being so overwhelmed was then.

It wasn't the math at all. I couldn't say, "What's the matter with mother? Why is she in the hospital?" But it was okay to say, "I'm upset about math."

Once I thought about that for a while and really believed it, I could read that stat. My head stopped hurting and I was doing fine. I just didn't feel hung-up anymore.

Ruby was able to connect her feelings about math to specific events in her family that might have seemed unconnected to math. Others have found that their reactions to math are related to family problems involving moving, separation, and divorce. For some it has more to do with traumatic school experiences. And sometimes math is felt to be very closely related to money:

Martha: Math and money are the same thing as far as I'm concerned. This is true with every aspect of money: making it, knowing math in order to be able to make money, going to a restaurant and computing a bill, or computing a budget.

Mike: When I was in the ninth grade, I didn't want to grow up and take responsibility. I still don't want to deal with money. I freeze up when I have to deal with it.

Have you had any insights into what special meanings math has for you? If you have, it is a good idea to discuss them with someone you are close to.

Not everyone finds dramatic connections between math and other events in their lives. No single event or experience may be connected to present feelings about math. It can be a case of having had a long history of difficulty with math and a feeling that it is something you can't do. These feelings are often related to some of the causes of math anxiety mentioned in the first chapter.

ENDING MATH AVOIDANCE

If you are good at something and like it, you tend to condition yourself to go toward it. You find areas where you can perform well. At the same time, you avoid things like math that give you difficulty or cause you to get anxious. Math becomes like a foreign territory and you never develop the skills that are required to be an inhabitant of that land. This is compounded by anxiety which can keep you from using the skills, like intuition, that you use in everything else.

Skills become less and less available to you when you don't use them. But that does not mean that they cannot be relearned.

> Fran: It's really fun to see it come creeping back. Like in setting up the Sherlock Holmes problem tonight and then saying, "Gee, I don't know how to get all those x's on the same side, so I can't find the answer." But it was still very reassuring to be able to get as far as I did. The old skills started creeping out of their box.

Those who like to write tend to do a lot of it. It may be just writing letters or keeping a diary. But that means developing skill at assembling thoughts and expressing them in writing. If you think you can't write, you may avoid it whenever possible. When you are called upon to do some writing it seems like a very difficult task. It is not surprising that your mind goes blank. You simply have no experience putting your thoughts on paper.

If you are comfortable with math you seek opportunities to use it, rather than to avoid it. So you provide yourself ample opportunities to practice. Then when you are called upon to do something that involves math, it feels natural rather than impossible.

Jessica: I asked my daughter to teach me fractions. She said, "Why didn't you ever ask me before?" I don't know why I didn't, it just never occurred to me.

Hal: There are these rulers with fractions and decimals on them. I never used to even look at them. It is something that used to bother me. Now I find that I am looking at them all the time.

Phil: I have a slide rule calculator and instruction book. I actually started to read the directions last weekend. I had never been able to look at the book. This time I was actually reading it and understanding it. I found myself doing a problem without even having to look at the example that was given.

Martha: I went out to lunch with someone and there was a very complicated check. Ordinarily, I would have refused to look at it. This time I figured it out in seconds.

Changes like these are of very great significance. They mark a turning point where math is approached rather than avoided.

LEARNING MATH

Think about the things you do best. Are they always easy? What do you do when you run into difficulty?

Henry: What I do best is drawing. I like to draw live models and study the body.

Stan: Do you ever have trouble getting a picture to come out the way you want it to?

Henry: Many times!

Stan: What do you do then?

Henry: I work at it. But I don't have the patience to work at a math problem.

Suppose you feel that you cannot draw. Your first reaction on being taught might be embarrassment about anyone seeing the immature quality of your work. Your hands would feel clumsy and their motions would most likely be rigid. Intellectually you would feel you should be able to learn the fundamentals but your tension would get in the way. Each difficulty would reaffirm your inabilities and inadequacies. You might end up saying to yourself, "Why do I need this anyway? I'm sure I can get through life without it."

There is a difference between not having the ability to do something and not knowing how to learn it. Usually the things you have excelled at are things you know how to learn. As a result of our workshops, we are convinced that difficulties with math have a lot to do with not knowing how to learn it. In Part IV, the emphasis will be on how to learn math.

If you feel that you have never mastered arithmetic, if decimals, fractions, and percentages drive you mad, if "x" stands for a mystery you've never been able to unravel, then the thought of learning enough math to meet your present goals can be intimidating. The feeling is that there is just too much to assimilate.

It is not nearly as bad as it seems. A page or two a day isn't much, but it adds up quickly. Two pages a day comes to 60 a month which is 360 pages in six months. A lot of math is contained in that many pages.

Catching up on math is no worse than catching up on anything else. It is slow in the beginning, but as you lay a firm groundwork you will find that it goes faster and faster.

PART FOUR

doing MATH

CHAPTER TWELVE

everyday math

We are going to present some more math material now. This is the same material that is given during the fourth Mind Over Math workshop. The focus remains the same—your reactions to math when you look at it and try to do it. There is no expectation that you will be completely free of anxiety.

Do you remember how strong your first reactions and anticipations were? Can you compare them to your present responses? Perhaps your anxiety is now turning more to curiosity.

What are your expectations at this point? Do you expect to be able to do this math? Do you care whether you can do it or not? Remember, this is not being set up as a test of yourself. Do you wish you didn't have to look at any more math?

Think about our style of presenting material. Do you like or dislike it? Does that affect the way you work with the material itself?

You will find many implied problems in the new material. Some of them you will be able to do, others you may not. Focus on what you can do rather than on what you can't. Write down any thoughts you have about the problem while you are doing it. These thoughts may be intuitive ideas and feelings.

What do you think of us for giving this problem? What do

you think of each section? Do you have only one response, or a variety of responses depending on the part you are looking at?

Get yourself a cup of coffee or a snack. Do as much as you can, but don't rush. Give yourself at least ten minutes. Don't hesitate to take more time. There is no advantage to getting done quickly. Remember that coming back to things you can't do will often give you new insights.

We had been trying to think of a problem in comparative shopping. But the problem with the problem was how to make a can of tomatoes interesting. We gave up.
Eight 4 oz. bottles of grape juice for $1.69.
A 12 oz. bottle of grape juice for a half dollar.
An 8 oz. bottle of grape juice for 33¢.

What would you think of someone who asked you a question like—which of the following numbers are the same?

$$3/5, \qquad 60, \qquad .6, \qquad .600,$$

$$\frac{1/4}{5/12} \qquad 9/10 \times 2/3$$

Dynamints cost 25¢ for .55 oz.
Tictacs cost 29¢ for ½ oz.

Since "x" is so mysterious maybe we should use "m" for mystery. Do you think you can solve any of these mysteries?

$m + 3 = 5 \qquad 3m = 15 \qquad 4m - 3 = 2m + 5$

Have you been having any internal dialogues with yourself while looking at the problems? What have you been saying?

Are there any parts of the material that are unclear? Do you wish that you could ask a question?

We would like to clarify any misunderstandings about what you are supposed to be doing. In the first problem, the idea is to decide which juice offers the best buy. In the second, you are to comment about the thoughts you have about people who ask this kind of question. Then try to decide which of the numbers are the same. In the third section, which would you buy, the Dynamints or the Tictacs? In the last, we do not expect you to use algebra unless that is easy for you. Try to "guess" what the answer is. Look at the material again before reading on.

It is tempting to get into the details of how each of the problems is resolved. However, it is better to focus on your reactions first; once you start commenting on the math you start to forget about the feelings it evokes. The first step in any problem solving is always to take account of your feelings rather than trying to suppress them.

After about ten minutes we ask everyone in the workshop to discuss the thoughts they were having about the problems while they were working on them.

Noreen: I enjoyed doing the problems. I wasn't conscious of anyone else. I was just seeing whether I could do them.

Annie: My first feeling was "What do they want." I said, "Well, they probably want to know which is cheaper." I thought that if you had asked that it would have been easier.

Henry: For me it's time consuming. I have to double and triple check to make sure I'm right because I'm not sure of myself.

Sarah: I just decided I was going to do it. I didn't care if the answer was good. That was the attitude I took and the wall didn't come down.

Do you find that you feel more willing to try math problems now? Perhaps you still get upset initially, but if you take a few moments to calm yourself down you discover there is much that you can do.

Ethel: I found myself looking at $1.69 a hundred times and not taking it in. Then I glanced at the bottom of the page and thought, "Oh, no, I'm in algebra again." At first I really wanted to cry but then I tried to figure out the first one. I don't know if it's right but I did it.

Then I went on to the second and said, "I can't do this either." But then I started looking at it again and got an answer that seemed obvious to me after a few minutes.

I attempted the last one and then went back to the second. I remembered something from school and took a guess.

It is easy for something you can't do to catch your eye. Then your thoughts go to what you can't do rather than to what you can. When you are aware this is happening to you, look away and focus on a different part of the page.

Cynthia: On the second problem, I thought anyone who would ask a question like that was crazy. I don't know how to do fractions and decimals. I don't think I ever knew that.

The third problem I just couldn't do. The last really surprised me because I was able to do it. Instead of saying "x equals . . .," I just said, "What plus three equals five?"

Henry: On the "What do you think of someone . . . ," my answer was, "Why? Why do you want to know? What are you getting out of this?"

Jessica: I wouldn't want to say what I really thought of someone who would ask that.

Jeff: I got answers to all of them but went about it the hard way. I converted everything to decimals. I was uncomfortable because I knew it wasn't the right way. But it was the only way I could do it.

Rachel: I started with the first problem. I felt like I was in the supermarket and got really absorbed in it. In the second problem I said, "I know I'm not good at fractions." I took a stab at what I thought would be the right answer. Then I went on to the Dynamints and came back to work on the fractions. But I still had problems with it.

There is nothing wrong with not knowing something. It's important to be able to assess what you can do so that you separate the problems you know and don't know from each other. Otherwise, you see one mass of mathematics that becomes so formidable you can't bear to look at it. The more accurately you can determine what you need to learn, the easier it is to find the resources to learn it.

In doing problems it is essential that you start with what you do know. Then you will feel confident in your approach. Refinements and improved methods come much later. Don't put down your own approach. That is one of those self-defeating games you play on yourself.

Sometimes you anticipate failure even when you are successful.

Carol: I kept thinking, "this is easy," when I looked at the first one. I remembered learning it in the eighth grade. But then I thought, "The next step is the one they are going to lose me on." My feeling is that what I can do, anyone can do.

Noreen: I wanted to ask if these were terribly elementary so I could judge myself. I was afraid you would say, "Yes, they are terribly elementary." Then I'd think, "There, I am stupid again." So I won't ask you. But I will sulk in my closet.

Focusing on what you can't do or worrying about what level you are on takes all the pleasure away. When you look at half a glass of water, do you say it's half empty or half full? Think about what you have done, not about what you haven't done.

Ruby: When I like your jokes, I like the problems. I have a brand of grape juice that I like and really don't care what the price is. When I saw that problem I wrote, "Oh, God, I hate this." That was the problem I had the most trouble with. It wasn't fun, it wasn't cute, it was stupid. If I could avoid hating some of the problems so much I might do better.

We have tried to keep the tone of the material we present as light as possible. But perhaps you have a negative reaction to this approach. When the type of juice becomes a factor in how you deal with a problem, it indicates that you are too involved with the problem. As with all math, the ideal response is a neutral one. When you get involved in the story or in the comments that surround the math, you lose sight of what it is that you have to do. Your primary task is always information gathering, followed by working out whatever calculations are necessary.

> Mike: I figured out the price for each ounce in the first problem. With the Dynamints I used the same method as in the first but it got awful.
>
> When I heard Ethel talking about how well she was doing, I started getting angry at her. I was hoping she messed up somehow.

Some people find they cannot get anywhere with these problems. If that is the way you have been feeling, can you think of the reasons why? Were you still feeling tense? Were there things you just didn't know? Before you read on, take a look at the problems again. See if you can make more sense out of them than you did before. Ethel couldn't do anything at first. She just found herself "looking at $1.69 a hundred times and not taking it in." Little by little she discovered more and more things she could figure out. You can do the same.

LOOKING AT THE PROBLEMS

We will now start going through some of the problems. It is more important to show you a way of looking at them than it is to show you a way of solving them.

No two people would approach the problems in exactly the same way. We ourselves wouldn't necessarily do the problems in the way we are now describing if we looked at them at a different time.

Take as much emotional distance from these problems as you

can. Even take some physical distance. Try moving the paper away from you rather than holding it close.

The idea in the first problem is to decide which amount of grape juice you would buy. Stan will describe what he would do.

Stan: My first reaction takes into account the fact that I don't like to do arithmetic. I especially don't like to do division. I can tolerate multiplication a little better but I don't like it either. Because of these attitudes, my approach is to see if I can find a way to avoid doing arithmetic. I would rather spend time looking for ways to avoid division than to just do it. I would look at the first problem while holding it physically away from me.

I see eight 4-ounce bottles for $1.69 and say to myself that means altogether I have 32 ounces for that price. Notice I am not reading ahead. I am only trying to understand the information as I go along. I leave the first statement and don't try to do anything with it until I have looked at everything else.

Then I see 12 ounces for a half dollar and 8 ounces for 33¢. I start looking for some way to compare the different situations. I might repeat the numbers to myself several times because they have a tendency to slip away. I need to take them in and absorb them.

Eventually it hits me that 32 ounces is four times as much as 8 ounces. So I compare the first and the last. In the first one I have four times as much juice as in the last so it should cost four times as much. Four times as much as 33¢ is $1.32. But $1.32 is less than $1.69, so the last one is cheaper than the first.

So far I have compared only the first juice price and the last, and I know the last is a better buy. I still haven't done anything with the second. I now look for a way to decide whether the second or the third is the better buy. I stare at them for a while, trying to see if there is any way to compare 12 ounces and 8 ounces. There is no direct comparison, but after some time I realize that twice 12 gives 24 and three times 8 also gives 24. That gives me a common number of ounces for comparison.

Since 12 ounces cost 50¢, twice 12 or 24 ounces would cost twice 50¢ or $1.00. Likewise, since 8 ounces cost 33¢, 24 ounces would cost three times as much or 99¢. So the third price is the cheapest—but only by a very little bit. In actual practice, you would not care about such a small difference.

This is just one of many ways in which this problem could be done. There is no preferred way, just the way *you* prefer.

The approach to the problem that was just described may seem like a shortcut. It is not. What takes time is thinking about what to do. It may seem clear afterward, but it takes awhile to see the relationships just described. This may actually take longer than a more direct approach.

The point of this discussion was to illustrate how Stan's feelings about arithmetic led him to approach the problem. In many problems this kind of comparison approach may not be possible because the numbers involved cannot be easily compared.

How you approach a problem always depends on your reaction to it. Just doing what you know well often feels better because you feel secure in that. You may prefer to find price per ounce because it is a method you can count on. That is a good enough reason to use it. There is satisfaction in whatever method you use.

If you are unsure how to compute price per ounce, change one of the problems to a simpler one that you can do intuitively. For example, if it had been 10 ounces for 50¢, how much do you think one ounce would be? Is it clear to you that it would be 5¢ an ounce? If it is not clear, think about it again, saying to yourself, "10 ounces cost 50¢ so each one of those 10 ounces must have been 5¢." Don't hesitate to spend time re-reading this if you feel something is still unclear.

Now ask yourself what you intuitively did to get 5¢ per ounce. You must have divided the number of ounces, 10, into the price 50¢. So the way to get price per ounce is to divide the number of ounces into the price.

In the first case, you have 32 ounces for $1.69. Taking some scrap paper and dividing 32 into $1.69 would give a little over 5¢. The price per ounce in the second and third cases would come to $4\frac{1}{6}$¢ and $4\frac{1}{8}$¢ respectively. So the last is just a little cheaper.

How have you reacted to this discussion? Did you follow everything that was said? There is no reason why you should understand everything immediately. Were you unsure about whether $\frac{1}{6}$ or $\frac{1}{8}$ was smaller? It is possible that you also had some trouble with the decimals. Don't expect to know them now if you have had trouble with them for a long time. If you are interested in learning a little more about fractions and decimals, there is a discussion in Chapter 18.

If everything seems clear now, perhaps you are wondering why you didn't get it in the first place. Remember, problems often seem simple after the fact.

If you made a careless mistake somewhere, don't be too concerned about it. It is better to focus on what you understood rather than on where you were careless.

The purpose in having you look at this first problem was to help you gain further insight into your responses and to underscore the way they guide your approach to math. If you also learn something about solving other problems of this type, that will be a bonus.

The second problem is a way of finding out what you do and don't know about fractions and decimals. All the expressions in the group turn out to have the same value except 60.

If you don't know decimals, fractions, or multiplication and division of fractions, then you should not expect to be able to do it. Just say to yourself, "I don't know how to divide or multiply fractions. Once I learn, I will be able to do the problem." It is much better to say, "I have to learn math," than to say, "I can't do math."

In the part about the cost of Dynamints and Tictacs you could again try to compute the cost per ounce. But it gets messy because the numbers are not so easy to work with. One difficulty with this problem may have to do with being unsure about decimals. You have to know something about decimals to compare the ounces. The fraction $\frac{1}{2}$ is the same in value as the fraction $\frac{50}{100}$ which is .50 when written as a decimal.

Now look at the problem again and try to do it intuitively:

Dynamints cost 25¢ for .55 oz.
Tictacs cost 29¢ for .50 oz.

Do you notice anything about the relative costs and quantities? After some time you would see that with Dynamints you are getting more (.55 is more than .50) than with Tictacs, and the cost is less. So if the price were the determining factor, you would buy the Dynamints.

In the last problem, the group of equations was meant to be solved intuitively. We did not expect you to know any algebra. Before reading on, look at the problems again. Can you "see" what any of the values of m would be? The third problem is more difficult to "see" and you may have to make some trials for the answer.

The first equation was $m + 3 = 5$. Is it apparent to you that m would be 2? Then you have $2 + 3 = 5$. That is the way you check your answer.

The second equation was $3m = 15$. This means that you have three m's and together they amount to 15. In this case m is 5 because three five's make 15.

The third equation was $4m - 3 = 2m + 5$. The more information there is in a problem, the harder it becomes to "guess" what the answer is. There are systematic ways of solving problems like this, but it is also possible to do it by guessing numbers and trying them.

What would your first guess be? Let's suppose you guessed five. You could try it and see if it works. Four m's would be 20; less 3 would be 17.

On the other side of the equation you have two m's plus five. If m were 5 that would be 10 plus 5 or 15. Since 15 is not the same as 17, 5 does not work.

What would you try next? If you try 4 you will see that that works.

When you first look at problems they often seem intimidating. You feel you just don't know enough to do them. If you knew algebra, you might have solved the last problem directly. But it is important to realize that you can solve more problems than you think just by using your common sense and intuition.

CHAPTER THIRTEEN

being yourself while doing math

What's the best learning situation for you?

If you are a visual person, you are probably most comfortable with pictures and diagrams. You enjoy taking in impressions of things. You prefer to see a lot of illustrations and find it difficult to absorb information you can't see.

If you are a verbal person, you are likely to prefer long expositions and discussions. Teachers who give long expositions and books that use more words than symbols are most helpful to you.

You may be just the opposite and find it difficult to follow another's exposition. You are most effective when you are working quietly, on your own, at your own pace. It may be easier for you to express yourself in writing than orally.

You may prefer working with others and find working alone difficult and uncomfortable. When you work with people, your ideas flow and you find it easier to grasp new ideas.

Awareness of your own learning style is important. You should always try to work in the way that is best for you. Working in a style that is not of your choosing is bound to make learning more difficult.

Since math requires very intense concentration, it is particularly vulnerable to outside stress. As we've already warned you, don't be surprised if you have trouble doing math when personal problems are on your mind.

Stan: I recall that after a calculus class that I had just taught, a student came up to me to ask about one of the homework problems. I had done the problem many times before but when I looked at it, it made absolutely no sense to me. I couldn't get anywhere with it.

I kept thinking about the doctor's appointment I had scheduled for later that afternoon. I had gnawing feelings of uncertainty and doubt. I just couldn't concentrate on the math problem.

I told the student I was unable to work the problem out and asked her to see me the following day. That afternoon, I kept my appointment and found out there was nothing seriously wrong. Then I began to think about the problem again. It was suddenly very clear and only took a few minutes to work out.

The environment you are working in can also have a marked effect on your ability to do math. Try to do math when there are a minimum number of distractions. Find a quiet place to work if that helps you to concentrate. Listen to music if that helps.

When you look at a problem, find your own way to do it. When you have a strong feeling about the answer, go with it. Say to yourself, "I feel sure this is right and I am going to work on proving it."

Learning mathematics requires coming to the point where concepts feel right and make sense to you. There is no way to rush this. It may take a long time to fully grasp a new concept but when you get it, it feels like the light has just come on.

TAKING BREAKS

Loretta: When I write I work with enormous intensity mentally and I can feel the discomfort physically. I am so oriented to trying to explain something that I can only work with that intensity for a paragraph at a time. Then I have to pull away from it.

I will dust the house or clean the closets. Then I go back to my writing and the phraseology becomes clear.

It is the same with math. When you feel you are going in circles or can't think clearly anymore, you have to take a break. This may mean working on another problem or doing some other work. You may want to lie down, go for a walk, clean your house, make a phone call, or have cup of coffee. Do whatever you feel like.

You are not stopping because you are giving up but because you need to take some distance from what you are doing. It is easy to get so immersed in a problem that you really lose sight of the point of it. Relaxation enables your mind to keep working unconsciously and to generate new intuitive ideas. You should make conscious use of this process in doing math.

SLOWING YOURSELF DOWN

Math must be done slowly and carefully. You need to read each word in a problem, think about it, and write down whatever is important.

You have to learn at your own speed. After you have learned how to do a problem, do it again to reinforce your understanding. Then do many more problems which are of a similar type. The first five or six problems always take time and may be frustrating. But then you'll begin to pick up speed. Eventually, you'll have done so many, you'll just look at the problem and know what to do.

You can probably sense when you are going beyond your own pace. If you try to type too fast, you will find the number of errors will go up sharply. You actually save time by going slower because finding and correcting errors is such a lengthy process. Calculations like multiplication and division have to be done particularly carefully to reduce the chance of errors.

Math is meant to be done with pencil and paper. Quick mental calculations are the result of practice and experience. They have little to do with ability. Some people have more interest in and capacity for mental arithmetic than others. In speaking with mathematicians, we have found that they are about equally divided between those who can do mental arithmetic quickly and accurately and those who cannot.

Until you make the decision to take the time you need, you will be doing battle against yourself, creating pressure that will interfere with learning. Once you make up your mind to do what you have to do no matter how long it takes, you will find it goes faster than you would have expected.

DAILY PRACTICE

When you have not used math for a long time, you get rusty. It is like doing crossword puzzles. If you rarely do them, they seem very difficult. When you do them all the time, you get to the point where you can do the one in the Sunday *Times* fairly quickly.

When you do math on a regular basis, you get used to seeing the difficulties that are inherent in math problems. You become accustomed to their wording. Even when you can't do a problem immediately, you become used to leaving a problem and coming back to it repeatedly until you can solve it. You don't get discouraged when you get stuck.

If you have to prepare for aptitude tests like the college boards or the business boards, we recommend doing math every day. It is like trying to keep physically fit. Plan on spending at least an hour a day, and don't skip any days. It is better to spend fifteen minutes than to spend no time at all. When you skip one day, it is hard to get yourself going again.

Learning goes on between your daily work sessions. It means that you are giving your unconscious a chance to work on problems and to absorb concepts. What is unclear one day may be completely clear the next.

AIDS TO DOING MATH

There is nothing wrong with using a calculator as an aid. It is fast and accurate and saves you the trouble of having to go through annoying computations.

We don't recommend doing anything on a calculator that

you have not first learned to do by hand. Using a calculator feels bad when you are not confident you could do the work yourself if you had to. Then you hit the buttons and pray the answer comes out right. But if you are sure you can do the work yourself, it is well worth saving time by doing it on a calculator. Then you will regard the calculator as an aid rather than a crutch.

Did you ever notice that in stores they have lists which give the tax to be charged at each price? It is a good idea to take some of the pressure off yourself by writing down facts like these that you particularly need to know.

If you need percentages in your work, be sure to have some of the conversions from percentages to fractions and decimals available to you. If you play golf and have trouble with the scoring system, keep some notes on a card.

We both learned the multiplication tables through repeated use rather than conscious memorization. We are opposed to the use of flash cards for learning multiplication tables, because it emphasizes rote learning. It gives the false impression that math has to be done fast and requires memorization.

Being comfortable about looking up facts takes the pressure off trying to retain information, and when the pressure is off you will find yourself retaining more.

CHAPTER FOURTEEN

doing math under pressure

Have you ever noticed a change in your tennis game when you switch from volleying to keeping score?

> Nancy: Under pressure I can't do anything. The wall comes down and I can't think. I know that when I am under pressure, I am not going to be able to figure anything out.

Under the pressure of being observed or measured it is easy to start losing confidence. This may show up in a poorer performance than you might otherwise be capable of.

Excessive pressure can make even the most competent person do badly. This is clearly seen in the errors made in such high stakes events as the World Series or Super Bowl games.

The first step in dealing with pressure is to assess your situation realistically. You have to distinguish between real and imagined pressures. Since the pressures you once experienced in school, on tests and from your parents, have become internalized, you may feel unwarranted tension in a new situation.

Did you experience the math material we have presented as a test? Many people refer to it as a test and feel disturbed because they didn't have enough time to finish or didn't do it fast enough. Yet the only purpose in looking at material was to provide a vehicle for bringing out and discussing reactions to math.

When you get a bill in a store, who says you have to be able to check it quickly? Why is it wrong to use a pencil if that helps you? Do you really care if it is right to the last penny or can you be content with its being "more or less" right?

When others are watching, you tend to project your expectations of yourself onto them. You feel sure they are thinking about how fast you are working and about whether or not you are able to do math in your head.

The more comfortable you are with doing math your own way, the easier it becomes for you to make the space you need for yourself. You can say, "I can't do this in my head," without feeling there is something wrong with saying that. You will also be able to say, "I am going to need some time to work this out. I can't do it right this minute." When you are able to do these things, you will begin to notice that others do the same.

> Jenny: My brother took us out for dinner the other night. It amazed me to see him take out a piece of paper to figure out the bill when it came. It didn't bother him. He's always been very good at math.
>
> I would never have done that. I would have been ashamed to take out a pencil and paper in front of everybody.

The more confident you feel, the less sensitive you are to others' opinions of you. It is easier to admit to mistakes and difficulties in areas where you feel secure than in areas where you do not.

PERSONAL STYLES

Are you a procrastinator? Do you leave everything to the last minute and then work yourself into a frenzy trying to get everything done?

If you work best under pressure, then putting things off to the last minute may be aggravating but not harmful. The last minute pressure of a deadline may actually help you work more efficiently. The problem comes when you procrastinate over something you are insecure about. Then the same pressures that mobilize you in other circumstances become overwhelming.

Most people who have trouble with math tend to put it off. When they find there isn't enough time, they get blocked because they have to work faster than they possibly can.

If you know you generally do not work well under stress, take special care to avoid putting yourself in a position where you have to do math under pressure. Allow plenty of time for whatever math you have to do.

DOING MATH IN PUBLIC

Sometimes there is no way to avoid doing math in front of other people. This happens in stores and at business meetings.

It is impossible to concentrate on whatever calculations you are trying to do, if you are thinking about what someone else is thinking. You have to remove yourself from the situation either physically or emotionally.

Peter: A few days ago, I took two people to lunch and I was expected to pay the bill. They were both very good at math.

> When the bill came, I had to add it up and also add the tip. They both sat there and watched.
>
> I froze and got panicky. The more I sat there, the harder they looked. I didn't know how to break that. Finally, I pushed my chair back from the table and put the thing in my lap so they couldn't see it. Eventually, they started to talk together and left me alone until I added it up.

In circumstances like the one above, it is perfectly appropriate to say something like, "I can't possibly do this with you watching me." Anyone can understand and accept the difficulty of working while being observed.

You can learn to consciously block out the other people. If you have some figures you have to look at or if there are some calculations you have to do, focus intently on the math. Decide to ignore the people around you by not looking at them and not listening to what they are saying. It is hard to do two things at the same time. You can't expect to be able to do any math if you are not giving it your full attention.

Ideally, you should avoid doing math in front of others unless it is absolutely necessary. Always consider the possibility of doing it later, by yourself.

TESTS

It is rare to feel completely calm when taking a test. That is just not likely to happen. In any kind of test situation there are the pressures of time and performance. It is natural to worry about how you are going to do.

As panic creeps up on you, it is easy to start getting nervous about being nervous. But nervousness on a test is quite natural. And it is possible to continue working despite some nervousness.

Stan: I always had a habit of panicking on math tests. Every time I would sit down to take a math test, I would immediately feel a flash of intense anxiety. I would sit there and really shake.

All I could think was "My God, I'm supposed to be a math major and they're going to find me out. I'm going to fail this test and they are going to ask what is wrong with me."

What I used to do was to give myself about ten minutes, on a one hour test, of what I called "panic time." I would actually allow myself to panic. I would sit there and say, "Well, here I go," and just feel the feelings.

I would watch the clock and say, "Only five minutes of panicking left." After five more minutes of feeling terrible I would start to work.

It helps to stop fighting yourself. Otherwise, you have little conversations going on in your head that go something like:

"I'm not going to be able to do it."
"Well, why don't you start working on it."
"I can't work. I'm too nervous."
"You better get to work. You're losing time."

You go around and around in circles that lead nowhere. It is better to accept that you feel terrible and go with the feelings for a while. You will find that you begin to calm down.

Allow yourself a few minutes of "panic time." You may as

well accept that you are not going to be able to work for a little while at the beginning of the test. Instead of getting caught up in your emotions, you acknowledge them so that you can cast them aside.

Taking a math test is just like taking any other kind. When you are ready to start working, look through the test for anything that you can do. Recall that with the material presented in Chapter 4, there was a tendency to be attracted to the hardest rather than the easiest problem. Always start with what you know, not with what you don't. It builds confidence.

If you find you can't do a problem after working on it for a little while, let it go and try another. You will often find that an idea about a problem you left will come to you while you are working on another or when you come back to the same problem later. It has the same effect as taking a break. Your unconscious continues to work on the problem while you are away from it. Sometimes there will actually be a clue in a later problem that will help you with one that you couldn't do.

It is a little like doing a crossword puzzle. You go through all the numbers down and across to see what you can figure out most quickly. That builds the supports for what is hardest for you to figure out.

In times of stress you almost instinctively retreat or regress to a previous secure position. You have to do something you are sure of before you can move forward again. Don't worry about using a method that seems elementary. Use your intuition. Count on your fingers. Try guessing numbers. Do it the "long" way. It doesn't matter how you begin, just as long as you do so.

Pressure may make you reluctant to give yourself the time you really need. But you can lose many valuable minutes constantly watching the clock during a test. Putting your watch away sometimes helps to take some of the time pressure off yourself. When you try to work faster than your capabilities permit, you will start making too many mistakes. It is better to stop counting the minutes and just work until you are told to stop.

If you look at the solution sheets for math tests, you will find that they are usually very short. That is because the actual writing time of a test is quite short. Thinking about the problems takes the most time. You need to allow yourself time to think. Part of that time can be devoted to allowing yourself to calm down.

> Arlene: When the paper is handed out and the pencil is in my hand and I see everybody doing something I think, "Oh, my God, they can all do it."

When you see your neighbor doing something, you naturally assume that he or she is doing what you can't do. That is not necessarily the case. In the Mind Over Math workshops, over half the people felt they were the only ones feeling anxious, confused, and unable to start.

On math tests many people start by recopying questions or writing down things that they will later discard. The paper that gets handed in first is usually from someone who has given up rather than from someone who has done everything well and gotten an A.

Whenever possible, you should ask questions. Your panic may be caused by misunderstanding or misreading.

There are few situations where questions are not permitted. Even on math tests, you can ask the instructor about anything that seems vague. The worst he or she can say is that they can't answer the question.

The only place where questions about a test are absolutely not allowed are on such standardized exams as the college boards. But they represent a very small portion of tests.

While there are very real pressures of time and performance on standardized tests, the same things apply to them that apply to tests and pressure situations in general. You can work only so fast without sacrificing accuracy. The most you can do about speed is to have practiced.

The tension of a test can make you lose confidence in your memory. If there are several formulas you have to know for a math test, and you are worried about forgetting them, look them over just before you go in. As soon as you get your test paper, write them down so they are there for you to refer to. Do this before you look at any questions. Writing them down immediately is not cheating and you will feel a little more relaxed knowing they are there.

Remember, tests are not the last word on your abilities. There is an art to taking tests, especially standardized ones. Many people with high scores in math on the college boards have trouble with it in college. And it is not unusual for someone who has done poorly in math on the boards to do very well at it in college.

The pressures of tests are very unnatural. As in most things in life, the important thing is what you accomplish, not how long it takes you.

CHAPTER FIFTEEN

reading a math book

Sharon: In a math course I took last summer, the first five or six pages were all right. But then they got into things that were beyond me. When we got up to page ten, I said forget it and went to the registrar to get my money back.

If a math book is at the right level for you, the first five or ten pages should be mostly understandable. You should be able to go through them without too much difficulty. After that there will be many topics that will be difficult. If that were not the case, it would mean you already knew the material and the book was at too low a level for you.

Math books build on themselves. The first few pages lay the groundwork for what is to be the real subject matter of the book. New ideas are confusing at first. This does not mean they are too hard for you. It just means the time has come to start reading carefully and work slowly.

The flow of a math book is not like the flow of a novel. A novel should read fluently, but math books do not. If you are reading a novel and are somewhat distracted, you can still get an idea of what it is about. You can skim and get a sense of it. When you are not fully concentrating on math, you will get very little out of it and it will seem more difficult than it really is. The expectation is that you will read a math book word by word and use pencil and paper while you are reading.

When you get to places where you have trouble, don't give up. Stop to figure out whatever you don't understand, be it a word, phrase, symbol, or example. That may require referring to another book on the same subject, going to a reference for background information, or just working with a pencil and paper.

Some pages will go very quickly, others very slowly. Progress through a math book is never steady and even. You will read along and be following, when all of a sudden there is something you don't grasp. What is missing is a kind of intuitive understanding. You may have to read, hear, or discuss some topics five or ten times before they become clear. Then something suddenly clicks and it becomes difficult to understand why you didn't understand before.

Math can be viewed as a succession of concepts, often interrelated. What may be easy for you will be hard for someone else and vice versa. When you master a concept, everything goes smoothly until you come to the next new idea. It is frustrating when you are struggling with a new concept and satisfying when you master it.

When you are learning to play a musical instrument, each new fingering is difficult and frustrating at first. It takes many hours of practice until it is mastered. Then you go on to the next new fingering and the process begins again.

Math books are usually not repetitive, so there is little chance of picking something up from reading on. Never start in the middle of a math book. It will be as confusing as starting with

French II rather than French I. Each page assumes you have mastered the previous page. Mastery can take minutes, hours, or days. But thoroughness pays off in the long run because it provides a strong background and eliminates the uncomfortable sense that you have missed something. Ultimately it enables you to move much faster.

Titles and introductions can be very intimidating. They often summarize all that is in a chapter and are harder to understand than the chapter itself.

> Ethel: My first look at the page was frightening. It said, PARABOLAS! My insides swelled and screamed, "What is a parabola? Oh, God, it's getting harder!"
>
> This was before I began reading. I flipped through the upcoming pages with dread. I sat there with my head in my hands and felt a headache coming on. All this occurred in about fifteen seconds.
>
> Nevertheless, I read on. Things didn't turn out half as bad as they had looked. I couldn't believe I was doing so well on something I had just dreaded.

USING MORE THAN ONE BOOK

When you don't have the background for a topic, it makes learning twice as difficult. You are trying to master new concepts when you lack the vocabulary which previous concepts provide.

For example, if you are trying to learn algebra and they talk about equations which have fractions in them, you will have extra difficulty if you were never comfortable with fractions. The best thing to do is to read about fractions in a book that reviews arithmetic.

Soft-covered books that review the concepts of arithmetic and algebra are available in most book stores. You should buy such books before you begin to study math. Use them for reference. The titles of these books often include the words "simple," "easy," or "elementary."

Titles of books are chosen to be encouraging and to make the book attractive. Don't take them too literally. Some topics are sure to be difficult.

You should have at least two books on whatever you are studying. There are a multitude of books on every topic in math and each has different discussions and examples. The same explanation is not good for everyone, but with two or three books you have two or three explanations of each idea available to you. It is likely you will find one that is helpful.

Advanced math books are very concise in their coverage of concepts from previous courses. If you are unfamiliar with a concept, brief discussions are difficult to understand.

When you read a more elementary book, with lengthy discussions, what had seemed incomprehensible becomes much clearer. Reading an elementary and advanced book simultaneously may be faster than working with the higher level book alone.

The choice of books is up to you. Pick books that appeal to you. If you are very verbal, a book with long explanations is likely to be most helpful. If you are very visual, you would probably choose a book that has more illustrations.

THE LANGUAGE OF MATH

It is easy to respond emotionally to math. The choice of words used is often very peculiar. For example, numbers are referred to as negative and positive, rational and irrational, real and imaginary.

If you think of the traditional meanings of these words, you might wonder who would want to deal with numbers that are negative or irrational. What sense is there in talking about a number that is not real but imaginary?

When you confuse the colloquial and mathematical usage of words, math seems very strange. Words need to be regarded impassively and taken only for what they mean in math.

In a foreign language, many words have the same spelling and a similar pronunciation to English words, but their meanings are entirely different. Those words, called false cognates, are the most confusing. For example, the Dutch word "monster" means sample. Likewise, the word "irrational" has nothing to do with sanity when used in math.

The words and symbols of math have very specific meanings. If you are at all unsure about the meaning of a term, look it up or ask someone to explain it.

Translation is also a key to doing word problems. As with the Sherlock Holmes story, these problems are meant to be analyzed word by word, not just read and solved. The important words should be written on scrap paper. When possible, words should be translated into mathematical symbols. Then, when you look over the problem, you won't have to take in a mass of confusing material, but can take in the major facts at a glance.

When looking at word problems, it doesn't matter whether they mention trains, cars, ditch diggers, or turtles. The words are there as a vehicle for presenting information. Your job is to sort out that information, to figure out the puzzle.

MATH BOOKS AND MATH CLASSES

Even when taking a math course with a good teacher, you have to be prepared to work on your own. Someone who understands math and knows how to present ideas well can make math feel comprehensible at the moment of presentation. Yet when you go home and try to do assigned problems, you may discover you can't do them.

You can't learn math just by listening. When you watch a pro play tennis it seems easy and you can see what you have to do. When you try to do it yourself, it is a different story entirely.

Problems look easy after they have been solved, because the difficult concepts and ideas are hidden in the final presentation. To make these concepts your own, you have to think about them yourself and try to apply them.

You will get the most out of a math class if you make it a practice to read over a topic before it is covered in class. You gain perspective on the material and clues about the areas where you will need to pay special attention to the instructor's explanations. After class, it is necessary to go over the material again, even if you felt you understood it in class.

Remember, you have the right to ask questions before, during, and after class. In colleges, instructors are required to schedule office hours. You should make an appointment to get answers to your questions. Most instructors are willing to help if you make your needs known. In fact, they are often disappointed when no students seek them out.

PART FIVE

CHAPTER SIXTEEN

feeling the difference

Our goal has been to help you overcome intense math anxiety so you can approach math with a positive rather than negative attitude. The fifth and last Mind Over Math workshop is devoted to summing up. We compare expectations with results and underscore what we think are the most significant changes. Small changes in long-standing patterns are always of great importance. What may seem like a little difference is often the most meaningful.

If we were given ten meetings to teach math to people who feared and disliked it, we would schedule no math for the first five. Instead, we would use the time to focus on feelings about math, as we have in this book. We would also develop a realistic view of what math is and how it is done.

At The Dalton School, a private school in New York City, we offered a series of workshops to several classes of ninth grade algebra students during five of their regular periods. No math was taught. Yet students' attitudes changed markedly and, in many cases, there were dramatic improvements in grades. Teachers found they were able to cover more material in two hours per week than they previously covered in three.

Little learning can take place in the face of intense anxiety.

When you are more relaxed and feel positively about math, you can read a book or listen to explanations without fear or panic. Then learning proceeds many times faster than if you were trying to fight off those feelings. You can learn more in five meetings when you feel calm than you can in ten when you feel anxious.

Think about what you expected to gain from this book. Where do you feel you have made changes? What have you found to be most satisfying?

> Fran: I wasn't expecting much. I just thought it was worth a try. Now I'm impatient to get going with some math.

It is a big change to find yourself looking forward to learning math and eager to get on with it. At the first exposure to material, with the story on pages 61–62 about what we did during the day, there was extreme anticipatory anxiety. Almost everyone would have been just as happy if there had been no math at all.

> Judy: Before I started I thought I was hopeless . . . too far gone. Now I feel that if I take my time and work slowly, I will be all right. I feel much better now. I'm not lost anymore.
>
> I tried to balance my checkbook. In the past, I always let my checking accounts go. At first it came out wrong, but I decided to take my time and figure it out and I found my mistake. It felt very good.

When you feel hopeless, it is easy to give up at the first sign of difficulty. As you gain self-confidence you will find yourself persevering until you succeed. Then you can build upon your successes and focus on what you can do rather than on what you can't.

Loretta: I have a heightened awareness of myself. I realize I am not as ignorant as I thought I was. There are specific areas I need to work on, but it is more a matter of learning more methodology than of a lack of aptitude.

Rachel: It is not a mortal sin if I don't understand fractions. It is a stumbling block I have to get over.

Cynthia: I realize I'm not dumb. I just don't have the tools. I have gained the confidence to get them.

NEW ASSESSMENTS

When you are anxious about math, it all looks like a blur. With decreasing anxiety you can realistically appraise exactly what you need to learn. It is no longer overwhelming.

Since math builds on itself, it is important to develop a strong base of knowledge. We recommend that you begin with a review of arithmetic unless you feel completely comfortable with decimals, fractions, and percentages.

The necessity for understanding previous topics is sometimes subtle. It may seem that there is only a single word or phrase you don't understand. Yet this word may have a very special meaning that is part of a concept you mislearned in a previous course. For example, when you take statistics they use a lot of formulas involving letters. If the concepts of algebra are unclear to you, then these formulas will give you trouble because it is in algebra that you first use them.

Starting from the beginning is not nearly so bad as it might seem. Suppose you were to start with a book that reviewed all of arithmetic. You would find that you know far more than you thought. By going through such a review, you would firm up what you know and fill in the gaps in your knowledge that have been troublesome for so long. You would gain confidence in yourself as you develop the foundation you need.

Ruby: I got an arithmetic review book and started going through it as a way of seeing how I was doing. I discovered that I was really not afraid anymore. I was not anxious.

I realized it was okay to go back and learn things all over again. When I started working, I started enjoying it. It was fun!

I thought I didn't know anything, but I found that what I didn't know was not as enormous as I had imagined. It was good to start with chapter one of the simplest book. It gave me a lot of self-confidence. Had I started in the middle of my statistics book, I would have felt deluged and never would have done it.

NEW PERSPECTIVES

What reactions have others had to your reading this book? Have they been encouraging? How has it affected your perceptions of yourself and others?

Fran: When I came home from the second session, my husband wanted to know what we were doing. I gave him the paper we had done. He dropped right into my ninth grade math teacher's role. He said, "Now, how did you get it? Oh, Fran, c'mon, you can figure that out."

Rachel: Last weekend, my husband gave me some problems. I couldn't do them. He said, "This is *the* method. You can't go off it."

If he had done that a month ago, I would have been very upset and felt very stupid. I didn't feel that way this time. I said I was going to do it my way.

When you feel better about yourself, you are much less likely to let someone make you feel bad. Rather than being immobilized by negative remarks, you deal with them assertively.

Jessica: Recently, my husband had to mix some paints in varying proportions. He explained the way he was going to do it and I said, "No, it's not going to work out that way." He said, "How do you know? You can't do arithmetic!"

I told him what he could do with the whole thing. I didn't talk to him for three days. Then he apologized.

Before, I would have felt badly. I would have said, "Well, you're right. I guess I can't do math." This time I *knew* he was wrong. I was absolutely furious, which is very different than my normal reaction which would have been to just laugh it off.

You convey how you feel about yourself in the image you present to others. When you think well of yourself, others will treat you well too. With decreased anxiety and improved self-confidence, you are able to appraise both yourself and others more objectively.

Peter: I have found it very liberating. I don't have to hide in the closet anymore. Hearing about others' difficulties has put a perspective on my own. I see myself as being a point on a spectrum of difficulties. I don't feel I must cope alone.

I learned that I have to give the same commitment to the learning process in math that I would with anything else.

Annie: Last week, a good friend of mine came over for dinner. I had always thought she was very good at math. I held up a chart for her to look at and she physically flinched. She said, "Don't show that to me. A wall comes down. I can't look at it."

You may have discovered that what you do with math, you also do in other areas. Once you understand your responses, you gain the power to control them. You can catch and stop yourself. Without this awareness, you are drawn into diversions without ever realizing what is happening to you.

Loretta: I came to realize that I have a tendency to go off on tangents. I become very involved in an irrelevant part of a problem. This is true of everything I do. But I am more able to bring myself back now.

Marilyn: It is a tremendous change for me to just be participating in a math anxiety workshop. I had never realized how much I tuned out with math and refused to deal with it. I also found out that I get frustrated and give up with anything I find difficult, not just math.

It is a big change for people to take a math anxiety workshop and deal with their problems head on instead of by avoidance.

It is an equally big change for you to have read this book. It means you are determined to do something about math. You will not let it stand in your way anymore.

SEEING THE DIFFERENCE

When you learn a new word, it seems to appear everywhere. Likewise, when you see math realistically, a whole new world opens up.

Have there been any situations lately in which you have handled math in uncharacteristic ways? Is there anything you have done that you wouldn't normally do? Any such differences are important.

Sheila: My son brought a board game home. I hate those games and normally would have refused to look at it. This time I actually played it with him. It was still boring, but at least I did it.

Eileen: I found that the bank made a mistake of $508.98. I didn't say, "Oh, well, it must be my mistake." It was good for me to feel confident that it wasn't me but the bank.

Marie: I never counted change before. A veil would come over my eyes and I would just hope it was right. Now I find I am looking at change and bills and not panicking. It seems obvious. I don't even have to figure some things out.

Henry: I was sitting at dinner the other day and my sister was reading math problems from a popular magazine. I was the only one in the room who was getting them. I didn't get anxious, I just told them the answers. My father is good in math and he wasn't getting any of them.

Fran: I had a very different feeling about the budget sheets that were passed out at the board of trustees meeting last night. Before, I would always let that whole part of affairs glide by me. I relied on the men in the group who were financial people and knew the right questions to ask.

I would blank the whole thing out and just go, "Uh, huh, uh, huh." This time I was having fun looking for relationships in the figures. I really wanted to get at it, understand it, and ask questions about it.

Dan: For years I have been trying to learn simple multiplication. I learn it and then forget it. This past week I asked a cousin to teach me and it suddenly all made sense. I'm sure I know how to do it now.

Noreen: I find that I can talk to my broker now. Even though I don't understand everything, I feel more on top of it. I feel less inferior.

Wendy: I started on my math book. I bought it two years ago and this is the first time I looked at it. It's really fun.

Peggy: I had to check the payroll this week. Usually I have to do it a couple of times. This time it was just bing, bing, bing, all the way through.

These changes are significant. They should not go unnoticed. If, for the first time, you find that you are looking at money and bills, it means you are on the road to being able to handle your personal finances. If you are balancing your checkbook when you couldn't even look at it before, it means you have begun to relax—and you have learned it without instruction.

Reconciling your checkbook requires calmly sitting down and taking a close look at records of how much was deposited and how much was withdrawn. Most mistakes are in transposing a number from one page to the next or in incorrectly recording the amount of a check.

An important change is in just being able to look at figures wherever they appear. You may have found that you have begun looking at the little graphs that appear in newspapers and magazines. These graphs provide a pictorial summary of information. When you look at the details of graphs you will notice key words that explain what information is being presented.

If you can look at numbers calmly, you will be able to handle yourself differently and you are ready to learn whatever math you need.

Sometimes you do not need to be able to do anything more than just look at figures. All that may be required is concentration and common sense. When given a choice between an assistant manager and a manager position, you are more likely to try for manager when you know you can handle budgeting—which does not require that you learn much math, just that you be comfortable with it. You feel more secure in your work when you know you are not going to be panicked by numbers. It may even be easier for you to ask for a raise!

APPROACHING MATH

Julia: I looked at all the x's and y's on a chart and studied it. It finally made sense to me. I had the patience where I had blocked it out before.

Phil: Five or six years ago I just couldn't consider the MBA. Now I'm anxious to do the math it requires.

Walter: Last weekend I went to the store to buy a paper, and a set of books on elementary math caught my eye. Ordinarily, I would have kept going. This time something made me stop.

I picked up one of the books, looked at it, and thought, "This would be interesting to read." Before, just the thought of it would turn me off.

If I saw a problem where you had to add 6 and 4, I would not see the 6 and the 4. I would see the numbers and become traumatized. Suddenly, I'm picking up books I was terrified to pick up before. To me that's huge progress.

A move in the direction of math after a lifetime of avoidance really is "huge progress." When you start feeling you can learn math, you no longer have to make decisions on the basis of math avoidance. You can move in whatever direction is important to you.

Annie: I can't believe what's happened to me. I'm definitely going to business school. There is no doubt in my mind. I signed up this past week. I've been going through the GMAT book every night. I can't believe the amount of math I know.

It's not that I do all the problems right. I get stumped more frequently than on the verbal part. But I see that there are also a great number of words I can't match antonyms to or don't know analogies for. But I see that I use the same reasoning for them that I do for math.

It's weird. It's almost insidious. I'll look at a problem and some of the old feelings will come back. Then I'll think, "Look, you haven't done that in six years. You also don't remember French vocabulary."

I do part of the math and then go to the verbal section. If I don't know a word in the analogies I think, "Well, I just don't know it." It's not anything that scares me.

I've also done strange things like walking down the street and adding up license plates. I'm just going completely crazy with the amount of math I know and the way I can do it.

It feels like a major breakthrough. I don't feel the tension, fear, and anxiety. I still fear taking a test. But there is no time when I haven't felt that. I can live with that. I know I still need some tutoring in certain areas. That is something I'm going to have to work an awful lot on.

It's strange. I've been at the University and I've been talking with people on the faculty. One person who I wanted to write a letter of recommendation for me asked about my strengths and weaknesses. He said, "Well, do you have a problem with math." I said, "No!!!" It was fabulous. If anybody asked me that before, I would have gone into a long diatribe about math and said, "Yes, I can't do math."

what it means

Fran: I have a feeling of release and feel if I can unravel math, then, by God, there are not going to be those other times of just drawing a blank. I feel like I can go on and conquer the world.

When you begin to find you can do something you always thought impossible, you really do feel you can "conquer the world." If you always thought you couldn't do math, and now realize you do have that ability, then what about all the other things you thought you couldn't do?

Karen found herself climbing ladders, changing fuses, and even repairing her motion picture projector when it got into trouble. She found she could use the same intuition and common sense in math and mechanical things that she used in every other area of her life.

Barbara went to Spain to write a story. Before leaving she decided to bring along her 35 millimeter camera which she had never been able to use. She read the instructions, practiced on a few rolls of film, and took terrific pictures.

> Connie: I feel I have progressed as an individual because I have taken on something I didn't know I could do before. I'm no great mathematician but now I am willing to try things. I feel better about myself. I feel that even if I didn't understand something at first, I could go home and figure it out.

When you are willing to try new things and feel confidence in your capacities, then you no longer feel as dependent on others to do things for you. You become increasingly independent. When you are blocked in math you are also blocked in other areas you see as related to math, from mechanical tasks to being able to read and follow instructions.

You may start thinking now about new directions. It may be learning to play tennis or to fix a car. It may be seeking the highest position rather than the second highest. It may be taking a course, going to business school, or studying medicine. It may be anything you want.

How much math you will choose to learn depends on your needs. You may decide you want to start over from the beginning or you may feel that is not where you need to put in the time and effort required for mastery. But you can make these decisions from strength rather than weakness. That is a big difference.

We all find ways to make use of what we know and feel competent in. You will begin to find unexpected opportunities to use math on the job, in personal matters, or in choosing a career.

The last part of this book is devoted to discussing several mathematical topics. Our purpose is to give you a feeling for the ideas of fractions, decimals, percentages, metrics, algebra, and calculus. We are not trying to "teach" these topics, but rather to provide a new perspective which will clarify some of the traditional places where people have difficulties. If you are interested in going further, you will be well-prepared to take a math course or work with a math book.

You are now in a position to realistically appraise books, teachers, and the help others give you. This will prevent you from playing negative "math games" on yourself. Even if you do experience some anxiety you will know where it is coming from and you will have the ability to reduce it. If you should decide to take a math course, it will be helpful to reread this book, especially Parts III and IV.

A persistent criticism of math is that it is too abstract and unrelated to life. But you have seen that doing math is not so different from any other creative endeavor. You can use the same skills in math you already use in what you do best.

In life, abstraction plays an important role. Moving back from situations and events can bring a clarity that is difficult to find when you are in the middle. By being objective, we get a chance to observe and analyze. By moving away we, in effect, get closer.

Sheila: I realized that I would click off whenever I came to something that was difficult for me. Once I saw that, I found I was able to bring myself back. I felt very powerful when I realized I could move the switch myself. But it also means that I have to take responsibility and can't blame anyone else anymore.

Ruby: I had an academic self-image that I was a good student, *but* couldn't do math. That was a nice limiting factor. I would feel "I could do this but . . . I can't do math." It is a little

> scary now, because if I don't have the "but," there is nothing to hide behind.
>
> It changes the way I relate to other people, because now if they say to me, "Oh, you can't do that," I say, "Oh, but I can." I am going to have to teach people to treat me differently.

The most meaningful change of all is in seeing yourself differently and taking control of your own life. We hope you have been able to do that.

PART SIX

supplemental

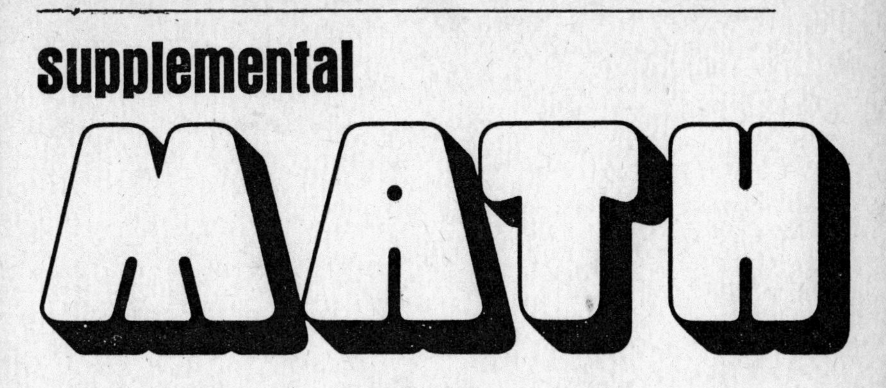

CHAPTER EIGHTEEN

fractions, decimals, and percentages

The purpose of this chapter is to give you a "feeling" for fractions, decimals, and percentages. The next step after reading this material is to study an arithmetic book. That will provide an opportunity to practice many different kinds of examples and consolidate what you have learned.

Learning about fractions requires starting with what you know and can visualize. Are you comfortable with halves and quarters? What do you think of when someone says one half? You may imagine the symbol $\frac{1}{2}$, half a pie, half a dollar, or anything else that is half of something.

When we see the fractions $\frac{1}{2}$ and $\frac{1}{4}$ we imagine half a pie or a quarter of a pie. Can you visualize one eighth of a pie? You can get one eighth by cutting a quarter in half.

Can you visualize one third of a pie? That is harder to see because it is more difficult to cut a pie into thirds than into quarters or eighths.

Fifths and sevenths are hard to visualize because it is hard to cut a pie into five or seven equal pieces. As you move from what you visualize easily to what you cannot see, you are taking a "leap of faith." You have to assume that sevenths are, in principle, very much like fourths. It is not necessary to visualize them exactly.

It is sufficient to understand that sevenths refer to the pieces you get when a pie is cut up into seven equal slices. Then just imagine that those pieces are smaller than the pieces you would get if you cut the pie into quarters.

Three sevenths of a pie would mean three of those things called sevenths. It is written $\frac{3}{7}$. Five twelfths, written $\frac{5}{12}$, means five of those things called twelfths.

The bottom of a fraction (called the denominator) always refers to the size (or denomination) of the pieces. The top (called the numerator) gives the number of pieces you have.

In comparing the fractions $\frac{1}{6}$ and $\frac{1}{8}$, the first is larger because when you cut a pie into six equal slices, the slices will be larger than if you cut the pie into eight equal slices.

ADDING FRACTIONS

Suppose you wanted to add $\frac{1}{8}$ and $\frac{3}{8}$. You could reason that if you had one of those things called eighths plus three more of those things called eighths, you would have four of those things called eighths, altogether. This would be written:

$$\frac{1}{8} + \frac{3}{8} = \frac{4}{8}.$$

You could further "reduce" $\frac{4}{8}$ by reasoning that four out of eight pieces means half the pie, so $\frac{4}{8} = \frac{1}{2}$.

How much would one quarter and one eighth be? Can you visualize this as three eighths? Draw some pictures to help you see this. The difficulty in adding comes from trying to add different things, namely quarters and eighths. You can reason that one quarter can be cut into two eighths and write:

$$\frac{1}{4} + \frac{1}{8} = \frac{2}{8} + \frac{1}{8} = \frac{3}{8}.$$

Now try to do $\frac{3}{4} + \frac{3}{8}$. First you convert the quarters to eighths. Each quarter is two eighths, so three quarters is six eighths. We can now write:

$$\frac{3}{4} + \frac{3}{8} = \frac{6}{8} + \frac{3}{8} = \frac{9}{8}.$$

Since eight eighths of a pie makes the whole pie, $\frac{9}{8}$ can be rewritten as $1\frac{1}{8}$. This means a whole and an additional $\frac{1}{8}$.

A more complicated situation arises when you try to add $\frac{1}{3} + \frac{1}{4}$. The difficulty is that quarters and thirds don't convert to each other. Look for another fraction they both can be converted into.

Thirds can be cut evenly into sixths, ninths, twelfths, fifteenths, etc. Quarters can be cut into eighths, twelfths, sixteenths, etc. Therefore, both quarters and thirds can be cut evenly into twelfths. We have $\frac{1}{3} = \frac{4}{12}$ (4 out of 12 is the same as 1 out of 3) and $\frac{1}{4} = \frac{3}{12}$ (3 out of 12 is the same as 1 out of 4). The solution to the problem is:

$$\frac{1}{3} + \frac{1}{4} = \frac{4}{12} + \frac{3}{12} = \frac{7}{12}.$$

Twelfths are what both quarters and thirds have in common, and 12 is called the common denominator.

How would you add $\frac{2}{3} + \frac{3}{4}$? Reason that each third gives four twelfths, so two thirds gives eight twelfths. Likewise, one quarter gives three twelfths, so three quarters gives nine twelfths. Now we have:

$$\frac{2}{3} + \frac{3}{4} = \frac{8}{12} + \frac{9}{12} = \frac{17}{12} = 1\frac{5}{12}.$$

Let's try one more example: $\frac{2}{5} + \frac{1}{3} = ?$ Try to do this before looking ahead.

The common denominator is 15 because both fifths and thirds can be divided into fifteenths: $\frac{2}{5} = \frac{6}{15}$, $\frac{1}{3} = \frac{5}{15}$. So,

$$\frac{2}{5} + \frac{1}{3} = \frac{6}{15} + \frac{5}{15} = \frac{11}{15}.$$

MULTIPLYING FRACTIONS

Can you visualize one half of one half? You may imagine cutting half a pie in half to get one quarter. We write this as:

$$\frac{1}{2} \text{ of } \frac{1}{2} = \frac{1}{4}.$$

By the same reasoning:

$$\frac{1}{2} \text{ of } \frac{1}{4} = \frac{1}{8}.$$

How would you figure out what half of three quarters would be? You could reason that half of each quarter is one eighth, so, if you have three quarters, half of that, would be three eighths. Then you would write:

$$\frac{1}{2} \text{ of } \frac{3}{4} = \frac{3}{8}.$$

When working with fractions, the translation of the word "of" is "times." Normally, when you think of multiplication you imagine things getting larger. When dealing with fractions, however, multiplication results in quantities getting smaller. So when working with fractions you say "multiply," but actually mean "take a fraction of." It is reasonable for quantities to get smaller when you take a fraction "of" them.

In the examples above, × can now be written instead of "of."

$$\frac{1}{2} \times \frac{1}{2} = \frac{1}{4}.$$

$$\frac{1}{2} \times \frac{1}{4} = \frac{1}{8}$$

$$\frac{1}{2} \times \frac{3}{4} = \frac{3}{8}.$$

When you look at these problems, to which we have reasoned out answers, do you see any relation between the left and right sides of each equation? Take a few minutes to do this. Looking for relationships is a major part of doing mathematics.

You will notice that, in each equation, if we multiply the tops (numerators) of the fractions on the left, we get the numerator on the right. If we multiply the bottoms (denominators) on the left, we obtain the denominator on the right. So *to "multiply" fractions, we multiply top by top and bottom by bottom.*

Try $\frac{1}{3} \times \frac{1}{4}$. Multiplying tops (1 × 1) and bottoms (3 × 4), we get:

$$\frac{1}{3} \times \frac{1}{4} = \frac{1}{12}.$$

Another example of the result of multiplying two fractions is:

$$\frac{2}{3} \times \frac{4}{5} = \frac{8}{15}.$$

Now we can do one of the fraction problems from Chapter 12.

$$\frac{9}{10} \times \frac{2}{3} = \frac{18}{30}.$$

You can reduce $\frac{18}{30}$ to $\frac{9}{15}$ (18 out of 30 is the same as 9 out of 15). Then $\frac{9}{15}$ can be reduced to $\frac{3}{5}$ (9 out of 15 is the same as 3 out of 5).

DIVIDING FRACTIONS

How many quarters are in one half? You can imagine half a pie being cut into two quarters and answer that there are two quarters in one half. The question just answered is written symbolically as:

$$\frac{\frac{1}{2}}{\frac{1}{4}} = 2.$$

In other words, that rather awful looking compound fraction (a fraction over a fraction) should be read as the question, "How many quarters are in one half?"

There are other ways to figure out the answer to this question. You could reason that there are four quarters in a whole pie, so there must be $\frac{1}{2}$ of 4 or two quarters in half a pie. We could summarize this reasoning as:

$$\frac{\frac{1}{2}}{\frac{1}{4}} = \frac{1}{2} \text{ of } 4 = 2.$$

Let's try another example. Think about how many eighths are in three fourths. Does your intuition tell you 6? You could reason that there are two eighths in each quarter, so there must be six eighths in three quarters (2 × 3).

You could also reason by the alternative approach suggested in the last problem. That is, in trying to decide how many eighths are in three quarters, you might imagine a whole pie as having eight eighths. Then three quarters of a pie must have $\frac{3}{4}$ of 8 eighths. That is 6. Summarizing, we have:

$$\frac{\frac{3}{4}}{\frac{1}{8}} = \frac{3}{4} \text{ of } 8 = 6.$$

How would you do this?

$$\frac{\frac{2}{3}}{\frac{1}{6}} = ?$$

The first step is to interpret the meaning of the compound fraction. It asks the question, "How many sixths are in $\frac{2}{3}$?" By the same reasoning as above, we could write:

$$\frac{\frac{2}{3}}{\frac{1}{6}} = \frac{2}{3} \text{ of } 6 = 4.$$

Since "of" translates to "times," we notice that, in each of the examples, there appears to be a relationship between division of fractions and multiplication. In the last example, 6 can be written as $\frac{6}{1}$, so it "looks like" a fraction. Now we can rewrite the solution to the last example as:

$$\frac{\frac{2}{3}}{\frac{1}{6}} = \frac{2}{3} \times \frac{6}{1} = \frac{12}{3} = 4.$$

We could also redo the first example in this section:

$$\frac{\frac{3}{4}}{\frac{1}{8}} = \frac{3}{4} \times \frac{8}{1} = \frac{24}{4} = 6.$$

You will notice that $\frac{8}{1}$ is the reverse (inverse) of $\frac{1}{8}$ just as $\frac{6}{1}$ is the reverse (inverse) of $\frac{1}{6}$. In both cases, a division problem has become a problem in multiplication. The second multiplier in this multiplication problem is the inverse of the fraction that was originally at the bottom of the compound fraction.

Summarizing, we say *to divide a fraction by a fraction, multi-*

ply the fraction in the numerator by the inverse of the fraction in the denominator.

Now we can do another of the fraction problems from Chapter 12.

$$\frac{\frac{1}{4}}{\frac{5}{12}} = \frac{1}{4} \times \frac{12}{5} = \frac{12}{20} = \frac{6}{10} = \frac{3}{5}.$$

The rule for dividing fractions applies no matter how bad the fractions look. This enables us to do problems that would be very hard to reason out or to do "intuitively."

Let's try one more example:

$$\frac{\frac{5}{6}}{\frac{3}{4}} = ?$$

See if you can do this before looking ahead to the solution.

The solution to this last problem is:

$$\frac{\frac{5}{6}}{\frac{3}{4}} = \frac{5}{6} \times \frac{4}{3} = \frac{20}{18} = 1\frac{2}{18} = 1\frac{1}{9}.$$

At this point, reading an arithmetic book on fractions should be much easier. It is not possible, however, to master fractions just by reading about them. Do lots of practice problems and reread this chapter and your arithmetic book whenever you experience conceptual difficulties.

DECIMALS

Decimals are fractions written in another way. The advantage of decimals is that they often make calculations easier.

The symbol .1 means $\frac{1}{10}$. So if you see .1, think one tenth. Likewise, .2 means $\frac{2}{10}$, .3 means $\frac{3}{10}$, and so on.

Here are some other decimal symbols, together with their meanings:

$$.01 = \frac{1}{100}$$

$$.001 = \frac{1}{1{,}000}$$

$$.0001 = \frac{1}{10{,}000}$$

Can you see the pattern? The number of positions to the right of the decimal point is the same as the number of zeros following the 1 in the denominator.

Some additional examples are:

$$.02 = \frac{2}{100}$$

$$.12 = \frac{12}{100}$$

$$.003 = \frac{3}{1{,}000}$$

$$.015 = \frac{15}{1{,}000}$$

$$.125 = \frac{125}{1,000}$$

The top of the fraction is always whatever number is written after the decimal point. The bottom is always 10, 100, 1,000, and so on, depending on the number of decimal positions. Take a few moments to study the patterns decimals follow.

How is .06 different from .60? We read the first as six hundredths ($\frac{6}{100}$) and the second as sixty hundredths ($\frac{60}{100}$). We have hundredths because in both cases there are two decimal positions. We could have written .06 as $\frac{06}{100}$ to emphasize that the numerator of the fraction is the number after the decimal point. Without a decimal point 06 is the same as 6, because zeros to the left of whole numbers don't change the value of the numbers. Zeros to the right of whole numbers do change their value. For example, 60 is quite different from 6.

Is there a difference in value between .6 and .60? To answer this question write each as a fraction.

$$.6 = \frac{6}{10}$$

$$.60 = \frac{60}{100}$$

But $\frac{60}{100}$ reduces to $\frac{6}{10}$, so

$$.60 = \frac{60}{100} = \frac{6}{10} = .6$$

How do .3, .03, .30, .003, .030, .300 compare? Write each of these decimal numbers as fractions:

$$.3 = \frac{3}{10}$$

$$.03 = \frac{3}{100}$$

$$.30 = \frac{30}{100} = \frac{3}{10}$$

$$.003 = \frac{3}{1{,}000}$$

$$.030 = \frac{30}{1{,}000} = \frac{3}{100}$$

$$.300 = \frac{300}{1{,}000} = \frac{3}{10}$$

You can now see that

$$.3 = .30 = .300$$

and

$$.03 = .030$$

while .003 is different from all the other numbers.

We can summarize by saying that *zeros to the right of decimal numbers do not change the value of the number.* Don't worry about remembering this. When there is the slightest doubt, carefully write the original decimal as a fraction. Then you will be able to "see" whether or not zeros are important.

The number 1.1 means 1 and .1 more; that is, 1 and $\frac{1}{10}$. It also can be written $1\frac{1}{10}$. As another example, 2.36 is 2 and .36 more or 2 and $\frac{36}{100}$. This is written $2\frac{36}{100}$.

CONVERTING FRACTIONS TO DECIMALS

It is easier to go from a decimal to a fraction than from a fraction to a decimal. Since decimals always refer to tenths, hundredths, thousandths, etc., converting from a fraction to a decimal requires you first to change the fraction to tenths, hundredths, and so on. For example,

$$\frac{1}{5} = \frac{2}{10} = .2$$

The fraction $\frac{1}{4}$ cannot be converted to tenths because 4 doesn't go into 10 evenly. But 4 does go into 100. So,

$$\frac{1}{4} = \frac{25}{100} = .25$$

Also,

$$\frac{3}{4} = \frac{75}{100} = .75$$

In the case of $\frac{1}{8}$, 8 divides evenly into 1,000 and

$$\frac{1}{8} = \frac{125}{1{,}000} = .125$$

It is helpful to have a procedure (algorithm) for converting fractions to decimals. In the case of $\frac{1}{8}$ the procedure looks like this:

```
   .125
8)1.000
  8
  20
  16
   40
   40
```

We have put a decimal point to the right of the 1 and added as many zeros as we needed to get an even answer (zeros to the right of a decimal don't change the value of the number to the left of the decimal). The decimal point in the answer is placed directly above the decimal point in the dividend.

Often fractions cannot be converted to decimals exactly. In that case we must "round off."

For example, to convert $\frac{2}{3}$ to a decimal you would start with

```
   .66
3)2.00
  18
   20
   18
    2
```

Can you see you would never get an exact answer? More sixes would appear if you kept on dividing. You can round off your answer to .67 since the remainder in the division (the 2) is more than half of the number you are dividing (the 3).

Arithmetic books discuss converting from fractions to decimals. To become proficient in this you first need to review division of whole numbers and then division of decimals.

MULTIPLYING DECIMALS

We discuss multiplication of decimals because that is useful in calculating percentages, which is the topic of the following section.

Suppose you wanted to multiply 2.31×1.7. One way to do this is to convert back to fractions. We write:

$$2.31 = 2\frac{31}{100} = \frac{231}{100}$$

($2 = \frac{200}{100}$ and an additional $\frac{31}{100}$ makes $\frac{231}{100}$), and

$$1.7 = 1\frac{7}{10} = \frac{17}{10}$$

Then we multiply:

$$2.31 \times 1.7 = \frac{231}{100} \times \frac{17}{10}$$

$$= \frac{231 \times 17}{100 \times 10}$$

We must work out the multiplication:

$$\begin{array}{r} 231 \\ \times\ 17 \\ \hline 1617 \\ 231 \\ \hline 3927 \end{array}$$

So,

$$2.31 \times 1.7 = \frac{3927}{1,000}$$

$$= 3\frac{927}{1,000} = 3.927$$

You will notice that to obtain the new numerator we had to multiply the original numbers without the decimal point. To obtain the new denominator we multiplied 100×10 to get 1,000. The answer was in thousandths and that gave us three places to the right of the decimal point.

Three represents the total number of decimal places in 2.31 (2 numbers to the right of the decimal point) and 1.7 (1 number to the right of the decimal point).

You could have arrived at the final answer directly by the following procedure: (1) multiply as if there were no decimals and (2) place the decimal point in your answer so that there are as many numbers to the right of the decimal point as in the total number of decimal places in the original numbers.

Let's do an example to illustrate this procedure; multiply 1.86×3.59.

$$
\begin{array}{r}
1.86 \\
\times\ 3.59 \\
\hline
1674 \\
930 \\
558 \\
\hline
6.6774
\end{array}
$$

The position of the decimal point in the answer was determined by adding 2 (decimal places in the first number) plus 2 (decimal places in the second number) to give 4 decimal places in the answer.

Mastery of decimal multiplication requires first reviewing ordinary multiplication. The only thing new in decimal multiplication is the proper counting of decimal places as just discussed.

PERCENTAGES

If you were offered a \$150 coat for 50% off, how much would you have to pay? Do you intuitively know you pay only \$75?

"Percent" always means hundredths. In other words, $50\% = \frac{50}{100}$. Since $\frac{50}{100} = \frac{1}{2}$, 50% off means the price is reduced by $\frac{1}{2}$.

Percentages can always be translated into fractions by writing the given percentage as the numerator and 100 as the denominator. For example,

$$3\% = \frac{3}{100}$$

$$10\% = \frac{10}{100} = \frac{1}{10}$$

$$20\% = \frac{20}{100} = \frac{2}{10} = \frac{1}{5}$$

$$25\% = \frac{25}{100} = \frac{1}{4}$$

$$75\% = \frac{75}{100} = \frac{3}{4}$$

Since percentages are really hundredths, they can also be directly expressed as decimals. For example,

$$5\% = \frac{5}{100} = .05$$

$$15\% = \frac{15}{100} = .15$$

$$30\% = \frac{30}{100} = .30$$

Working with percentages requires first converting the given percentage to a fraction or decimal.

COMPUTING PERCENTAGES

Suppose you had to compute 25% of \$80. To see what to do, first "translate" 25% to $\frac{25}{100} = \frac{1}{4}$. Then rewrite the problem "compute $\frac{1}{4}$ of \$80." Can you "see" that the answer is \$20?

Let's examine this problem again to gain additional insight into working with percentages. Recall that with fractions "of" translates to "times." So, "compute $\frac{1}{4}$ of \$80" is the same as saying $\frac{1}{4} \times \$80$. Summarizing, we have:

$$\begin{aligned} 25\% \text{ of } 80 &= \frac{1}{4} \text{ of } 80 \\ &= \frac{1}{4} \times 80 \\ &= \frac{1}{4} \times \frac{80}{1} \\ &= \frac{80}{4} = 20 \end{aligned}$$

As another example, compute 8% of \$16. The solution is:

$$\begin{aligned} 8\% \text{ of } 16 &= \frac{8}{100} \text{ of } 16 \\ &= \frac{8}{100} \times 16 \\ &= \frac{8}{100} \times \frac{16}{1} \\ &= \frac{128}{100} = 1\frac{28}{100} = 1.28 \end{aligned}$$

You can also use decimals to compute percents. In the last example, you could have written:

$$8\% \text{ of } 16 = \frac{8}{100} \times 16 = .08 \times 16$$

Then you would compute:

$$\begin{array}{r} 16 \\ \underline{\times .08} \\ 1.28 \end{array}$$

As another example, compute 12% of 85. The solution is:

$$\begin{aligned} 12\% \text{ of } 85 &= \frac{12}{100} \times 85 \\ &= .12 \times 85 \end{aligned}$$

You could now do the problem with the fractions or decimals. Using decimals, we have:

$$\begin{array}{r} 85 \\ \underline{\times .12} \\ 170 \\ \underline{85} \\ 10.20 \end{array}$$

So,

$$12\% \text{ of } 85 = 10.20$$

TIPS

The usual tip in restaurants is 15%. Using the methods of the last section, we could compute the tip on a bill of $17.60 as follows:

$$15\% \text{ of } 17.60 = \frac{15}{100} \times 17.60$$

$$= .15 \times 17.60$$

Multiplying, we get:

$$\begin{array}{r} 17.60 \\ \times\ .15 \\ \hline 8800 \\ 1760 \\ \hline 2.6400 \end{array}$$

So the exact 15% tip would be $2.64. In practice, of course, you would not worry about leaving exactly 15%.

You could compute an approximate tip by first "rounding off" the bill to $18.00 The total 15% tip can be viewed as 10% and 5% more. Notice that 5% is half of 10%. First we compute 10% ($\frac{10}{100} = \frac{1}{10}$) of 18:

$$\frac{1}{10} \times \frac{18}{1} = \frac{18}{10}$$

$$= 1\frac{8}{10} = 1.8 = \$1.80$$

Since 10% is $1.80, 5% is half that or $.90. So the total tip is $1.80 + $.90 or $2.70. That is surely close enough.

With a little practice you can use this scheme to compute tips in your head without pencil and paper.

What is the tax where you live? In New York City, the tax on all restaurant bills is 8%. This figure is written on the check. Notice that twice 8% is 16%. That is about the tip you would leave. So in New York City, you can compute the tip by doubling the tax. If your local tax were 7%, you could get the tip by doubling the tax (14%) and adding on a little bit more. Similar schemes can be devised with other taxes.

MARK-DOWNS

If the list price of a stereo were $350 and your dealer said he would give you 30% off, how much would you have to pay?

The phrase "30% off" means the dealer will take 30% of the $350 "off" before selling it to you.

$$\begin{aligned} 30\% \text{ of } \$350 &= \frac{30}{100} \times 350 \\ &= \frac{3}{10} \times \frac{350}{1} = \frac{1050}{10} \\ &= \$105 \end{aligned}$$

You could also figure that $\frac{1}{10}$ of $350 is $35, so $\frac{3}{10}$ would be 3 × $35 = $105. To arrive at the sale price you would subtract $105 from $350 to get $245.

Another way of doing this problem will save you a little

work. You could reason, if 30% of the $350 is taken off, then 70% remains. So *your* price is 70% of $350:

$$70\% \text{ of } \$350 = \frac{70}{100} \times 350$$

$$= \frac{7}{10} \times \frac{350}{1}$$

$$= \frac{2450}{10} = \$245$$

As before, you could have figured that $\frac{1}{10}$ of 350 is 35, so $\frac{7}{10}$ of 350 is 7 × 35 = $245.

In this problem, the numbers worked out easily because the given percentage is a multiple of 10. Let's try a problem where that is not so.

Suppose a $450 sofa was offered at 35% off. To determine the sale price, you could first reason that 35% off means you must pay 65% (100% − 35%) of the price.

$$65\% \text{ of } \$450 = \frac{65}{100} \times 450$$

$$= .65 \times 450$$

$$= \$292.50$$

Here we have omitted the work of multiplying .65 × 450; you should check to see that the result is correct. We chose to use decimals, but you could have stayed with the fractions.

At the end of Chapter 20, in which we discuss algebra, you will find additional examples of percentage problems. Knowing the sale price and the discount, you will be able to find the original price.

MARK-UPS

Merchants "mark up" their goods a certain percentage of the wholesale price to arrive at the retail price.

Suppose the wholesale price of a necklace is \$90 and it is marked up 80%. That means 80% of \$90 is added to the wholesale price to obtain the selling price. In summary, the retail price is:

$$\$90 \text{ (wholesale price)} + 80\% \text{ of } \$90 \text{ (mark up)}$$

We compute 80% of \$90 = \$72. Check the calculation yourself. So, the retail price is:

$$\$90 + \$72 = \$162.$$

As another example, try to compute the retail price of a typewriter which costs \$180 wholesale and is marked up 35%.

The solution is:

$$\begin{aligned} \text{Retail price} &= 180 + 35\% \text{ of } 180 \\ &= 180 + \left(\frac{35}{100} \times 180\right) \\ &= 180 + (.35 \times 180) \end{aligned}$$

$$= 180 + 63$$

$$= \$243$$

SUMMARY

In this chapter, the choice of topics was based on questions we have been asked most frequently. We have offered an "intuitive" approach to thinking about problems. If you have additional questions, we suggest you work with a book that reviews basic math or arithmetic.

mind over metric

Learning metric does not mean doing a lot of math or memorizing numerous conversion facts. It just involves getting used to some new expressions. This will happen naturally from daily use. We would like to give you an intuitive feeling for some of the metric quantities.

TEMPERATURE

Suppose we told you that when we went outside it was 20°C (C stands for Celsius). At first this doesn't tell you anything at all. However, if we added that it was very comfortable outside, neither hot nor cold, you might begin to get an idea of the weather. In fact, 20°C is the temperature that most of our homes are kept at. What would you guess that is in Fahrenheit (F)? From the description, your guess is likely to be somewhere between 68° and 72°F. Actually, 20°C = 68°F.

You need to know a few other facts about Celsius. Freezing is 0°C, while it is 32°F. Boiling is 100°C, compared with 212°F. Also, Celsius degrees are a little less than twice as large as

Fahrenheit degrees. The exact value is 1.8 times a Fahrenheit degree, but there is no reason to use this.

What do you think 30°C is in Fahrenheit? You could use 20°C = 68°F as a reference point. Then you could reason that 10°C is about 20°F, since a Celsius degree is about twice as large as a Fahrenheit degree. So the 30°C must be about 68° + 20° = 88°F. The exact conversion value is 86°F (68 + 1.8 × 10) but, for all practical purposes, 88°F is close enough. After all, could you really feel the difference between 88° and 86°?

After a short while of living with Celsius degrees, you would know what 20°C or 30°C felt like, as you now know what 68°F and 88°F feel like.

Let's try another example. What temperature do you think 5°C is? You might start by thinking that it is not too far from 0°C which is the freezing temperature. That compares with 32°F. A difference of 5°C is like a difference of about 10°F (twice as much) so 5°C would be about 32° + 10° = 42°F. The exact value is 41°F (32 + 1.8 × 5).

The nice thing about metric units, once you get used to them, is that they are divided up in a more natural way than the units we use. For example, doesn't it seem more reasonable to have freezing be 0° rather than 32° and boiling at 100° rather than 212°? Did you know that boiling was 212°? You will surely find it easier to remember boiling is 100°C.

LENGTH

Could you visualize exactly how far 100 yards is? That is very unlikely. Rather, you have a "feel" for how much distance it represents. A meter is a little bit bigger than a yard. If you see a sign that says 100 meters, it is enough to imagine about 100 yards. For your information, although you will probably never need it, one meter is 39.37 inches compared with 36 inches in a yard.

Yards and meters are divided into smaller units to enable us to make more accurate measurements of small items. Just as a yard is divided into 36 inches, a meter is divided into 100 units called centimeters (cm). "Centi-" means hundredth, so a centimeter is one hundredth of a meter.

Think about these divisions for a moment. If you had to design a measuring system, what would your choice of division be? Unless you were used to it, it is extremely unlikely that you would ever choose 36 as the number of divisions you were going to make. You would probably choose to divide a meter into 10 or 100 equal parts.

To compare inches to centimeters, you can think of 2 inches as being about 5 centimeters (the exact conversion is 1 in = 2.54 cm). About how tall are you in centimeters?

Suppose you were 5 ft 6 in. That is the same as being 66 in, since a foot is 12 inches. Every 2 inches converts to 5 centimeters, so your height in centimeters would be 33 × 5 or 165 cm. The exact conversion would be 167.64 cm (66 × 2.54). These little calculations should be done with pencil and paper. Don't try to do them in your head.

Although we offer the exact values for comparison, there is no reason to work them out. What is important is knowing approximately what the conversions are. When we finally convert to metric, you will not be busy calculating conversion values. You will get used to the new values so, for example, when you hear that someone is about 168 cm tall, you will imagine the same person you now imagine when someone mentions 5 feet 6 inches.

Let's try another example. Suppose you were 6 ft 1 in. About how tall would you be in centimeters? Six feet is 72 inches (6 × 12 in.), so 6 ft 1 in. is 73 inches. Every 2 inches converts to about 5 centimeters, so 72 inches is about 36 × 5 or 180 cm. We have not forgotten about the 73rd inch, we have just postponed it. Since 2 inches is 5 centimeters, 1 inch is about $2\frac{1}{2}$ centimeters. So 73 in. is about $182\frac{1}{2}$ cm. The exact conversion is 185.4 cm (73 × 2.54).

BIG DISTANCES

When we think of long distances we usually think of miles rather than inches, feet, or yards. Do you know how many feet or yards there are in a mile? Few people do.

A mile is 5,280 feet or 1,760 yards. These are not simple or obvious numbers and there is no reason why you should remember them. In the metric system, instead of talking about miles, we talk about kilometers (km). "Kilo-" means thousand, so kilometer means 1,000 meters. Meters and yards are about the same size, so a kilometer is quite a bit less than a mile (1,000 meters compared with 1,760 yards).

You'll get a good idea about distances if you think of 2 kilometers as being a bit over a mile. The precise conversion is 2 km = 1.24 mi.

Suppose you were driving on a freeway and saw a sign which said the speed limit was 100 km per hour. At first glance, this would look strange because you would tend to think of miles per hour.

Remembering that 2 kilometers is a little over one mile, you would think that 100 kilometers was a little over 50 miles. So you might guess the speed limit was somewhere between 55 and 60 miles per hour. For reference, the exact conversion is 62 miles per hour (2 km = 1.24 mi, so 100 km = 50 × 1.24 mi = 62 mi).

When there is conversion to metric, the speedometers will also be in metric. There will be no reason to do anything other than watch your speedometer.

WEIGHTS

Kilograms are a little over twice the size of pounds. In other words, 10 kilograms of potatoes would be more than 20 pounds. The exact conversion is 1 kilogram (kg) = 2.2 pounds (lb). So 10 kg = 22 lb.

For most purposes, in going from kilos (kilograms) to pounds, you will get a good enough idea of relative weights if you double the number of kilos to estimate the number of pounds.

If someone told you, on the phone, that he weighed 70 kg, how much would you estimate his weight to be in pounds? Double 70 would be 140 and, perhaps, you would throw in a little extra because you know a kilo is more than twice a pound. You might guess between 145 and 150 pounds. The exact value is 154 pounds (70 × 2.2). Although your guess would not be exact, it would give you a good idea of what the person looked like if you also knew his height.

Suppose you wanted to convert from pounds to kilos. For example, you might want to figure out your weight in kilos. Since 1 kg = 2 lb, approximately, $\frac{1}{2}$ kg = 1 lb (the exact value is $\frac{1}{2.2}$ kg = 1 lb). In other words, a pound is a little less than half a kilo.

If your weight is 120 pounds, how much do you weigh in kilos? You can halve the number of pounds to get 60 kg and take off a little bit since a pound is a little less than half a kilo. You might guess your weight to be about 55 kg. The exact value is 54.5 kg ($\frac{1}{2.2} \times 120$).

If you were shopping in kilos, you would halve what you order in pounds. Instead of 10 lb of potatoes, you would ask for 5 kg.

Both kilos and pounds are divided into smaller units. A pound is the same as 16 ounces and a kilo is the same as 1,000 grams. As you might imagine, a gram is a very small quantity. How much would you estimate 100 grams to be? You could reason that 500 grams is half a kilogram, about one pound. Then 100 grams would be about one fifth of a pound.

LIQUIDS

In metric, liquids are measured in liters instead of quarts. A liter is very close to a quart (1 liter = 1.06 quarts) and there will be no difficulty if you think of a liter as a large quart. Quarts are divided into 32 liquid ounces, while liters are divided into 100 centiliters. So 3 centiliters are about equivalent to 1 liquid ounce.

How much would 25 centiliters be? You could think of 25 centiliters as one fourth of a liter (25 out of 100 centiliters), or about one fourth of a quart, or one cup.

There will be little need to worry about changing back and forth between ounces and centiliters. Recipes and measuring cups will be in metric units and, if you still want to use an old recipe, you can still use your old measuring cup.

CHAPTER TWENTY

algebra

This chapter will present a new perspective on some of the main ideas that are traditionally included in algebra.

When you think of algebra, what probably comes to mind is the use of the letter "x." That letter becomes shrouded in mystery and quickly becomes the symbol of all things that are confusing. To demystify x, we will go back to the early grades when you worked with apples and oranges.

At first you may have been given problems like, "How much are three apples and five apples?" Then you would say, "Eight apples!"

$$3 \text{ apples} + 5 \text{ apples} = 8 \text{ apples}$$

The next thing to happen in school is that the apples are removed and you start writing $3 + 5 = 8$. This is a much higher level of abstraction because the objects of reference are gone. Very young children may be able to add three apples and five apples or three grapefruits and five grapefruits, but they may be totally unable to work out $3 + 5 = 8$.

To learn new mathematical ideas we first go back to what is simplest and best understood. This often provides insights that are helpful in solving harder problems.

Perhaps they never should have removed the apples. Then if you let "a" stand for apple you would be writing:

$$3a + 5a = 8a.$$

Likewise, if "p" stood for pear, you could write:

$$3p + 5p = 8p.$$

These abbreviations start looking more like algebra. You can visualize the mysterious x in the same way:

$$3x + 5x = 8x.$$

You may think of x as standing for x ray and read, "Three X rays plus five X rays is eight x rays."

When you look at the expression

$$2t + 7t = 9t$$

what do you think of? You could be letting t stand for anything; turkeys, for example. Would the equation still make sense if you let t stand for tens? Then you would read, "2 tens plus 7 tens is 9 tens." That is a true statement because 2 tens are 20, 7 tens are 70, and $20 + 70 = 90$, which is also 9 tens.

Let's try another example. Suppose t stood for twelves. Then the same equation would read, "2 twelves plus 7 twelves is 9 twelves." That is also true because 2 twelves are 24, and 7 twelves are 84, which comes to 108 altogether. That is the same as 9 twelves.

We are emphasizing that letters can stand for numbers as easily as they can stand for objects. This involves a conceptual leap basic to all of algebra. It means looking at the expression $2t + 7t = 9t$ in a new way. You can still visualize the letters as objects, but you have to accept that those objects can also be numbers.

FORMULAS

Once you get used to letters standing for numbers, it becomes easier to understand formulas. First, however, we need to say a little more about the language of algebra.

When we wrote 2t, we meant 2 t's whether the t's were turkeys or twelves. Whenever a number stands next to a letter it means multiplication. So 3a means 3 times a, 5p means 5 times p, and so on. When a letter stands next to a letter, it also means to multiply. So, for example, ab means a times b. If a stands for the value 5 and b stands for the value 6, then $ab = 5 \times 6 = 30$.

What is the value of $4a + 7$ when a is 3? The answer is 19. We got 19 by multiplying the value of a, or 3, by 4 and then adding the 7.

The formula for the area of a rectangle is:

$$A = LW$$

Here A stands for area, L for length, and W for width. We can read this formula as, "Area is length times width," since the placement of the letters L and W next to each other means multiplication.

The formula for the perimeter (the distance around the outside) of a rectangle is:

$$P = 2L + 2W$$

This formula can be read as, "The perimeter is twice the length of the rectangle plus twice the width."

If you were told that the length of a rectangle was 8 inches and the width was 7 inches, what would the perimeter be? It would be $2 \times 8 + 2 \times 7$, which is 30 inches.

As a final example of reading formulas, we will consider the formula for changing Celsius degrees to Fahrenheit degrees:

$$F = \frac{9}{5}C + 32$$

This says that to get Fahrenheit degrees, multiply the Celsius degrees by $\frac{9}{5}$ and add 32 (notice that 32 is freezing in Fahrenheit). For example, to change 20°C to Fahrenheit degrees, first multiply:

$$\frac{9}{5} \times 20 = \frac{9}{5} \times \frac{20}{1} = \frac{180}{5} = 36.$$

Then add 32:

$$36 + 32 = 68°F.$$

With practice on many examples, you will come to see formulas as abbreviated sentences. They are often derived by symbolically writing down what you already know, to offer a lot of information in a small amount of space.

EQUATIONS

You have already seen examples of equations in Chapter 12. When you think of an equation, imagine a seesaw in balance. For example, when looking at the equation

$$m + 3 = 5$$

think of the left side as balancing the right.

Solving an equation means finding out what value of m makes the balance perfect. The answer, in this case, is $m = 2$, because $2 + 3$ is exactly the same as 5. You wouldn't think of m being say, 4, because 4 would make the left side heavier than the right.

Many equations can be solved intuitively. That is, you can "see" what the answers are. For example, can you see what value of n balances the equation $3n = 9$? The answer is $n = 3$, because 3 times 3 is 9.

Try to see the answer to the equation:

$$3n + 2 = 14$$

Stare at this for a few moments. Experiment with some numbers. After some time you will come to the answer $n = 4$, because 3 times 4 plus 2 is fourteen.

As equations become increasingly complicated, it becomes harder to see the answers. Then methods of solving equations become helpful. If you view an equation as representing a balance between left and right, then the general principle of working with equations is to avoid upsetting this balance. This idea is expressed by saying, "*Whatever you do to one side of the equation, you must also do to the other.*" This will guarantee that the balance you start with is preserved.

In the equation $3n + 2 = 14$, it is a little hard to see the answer because a 2 is added to the 3n. We would have been happier if the 2 were with the fourteen. It is okay to take the 2 away from the left side if you also take it away from the right. This is written:

$$3n + 2 - 2 = 14 - 2$$

or

$$3n = 12$$

Now it is much easier to "see" that $n = 4$.

Let's consider the following equation:

$$5a = 2a + 6.$$

Can you see the answer? The difficulty here is that we have a's on both sides of the equation. It would easier if all the a's were together on the left side. We can remove the 2 a's from the right if we also remove 2 a's from the left. This is written:

$$5a - 2a = 2a + 6 - 2a.$$

The a's on the right can be grouped together and they cancel each other out. Also, 5 a's less 2 a's is 3 a's, so we have:

$$3a = 6.$$

Now you can see that $a = 2$.

Let's look at an equation from Chapter 12:

$$4m - 3 = 2m + 5.$$

This is the most difficult equation we have looked at because it has numbers and letters on both sides. It would have been easier to visualize if all the numbers were on one side and all the letters on the other.

The problem has to be done in two steps. First, concentrate on the numbers that are not connected to letters. Think about whether you would like to get rid of the -3 or the $+5$.

Suppose you decided to get rid of the -3. What would you have to do? Adding 3 to the left side cancels out the effect of subtracting 3 and leaves you with just 4m. But whatever is done to one side of the equation must also be done to the other, so we have:

$$4m - 3 + 3 = 2m + 5 + 3.$$

We have added 3 to both sides of the original equation and everything still balances. We are left with:

$$4m = 2m + 8.$$

The original equation has been transformed into a less complicated one. The 2 m's on the right can be removed by subtracting 2m from both sides of the equation. We then have:

$$4m - 2m = 2m + 8 - 2m$$

or

$$2m = 8.$$

Now you can see that $m = 4$, since twice 4 is 8.

The first time you work through a new type of problem always takes the longest. Once you get used to the reasoning and practice on many other problems, you will find it comes almost automatically. There will be nothing to memorize.

Let's do one more problem just like the last. We'll use different letters to emphasize that the choice of letter is really irrelevant. Try to solve the following:

$$5r - 2 = 3r + 4.$$

To get rid of the -2 on the left, we add 2 to both sides:

$$5r - 2 + 2 = 3r + 4 + 2$$
$$5r = 3r + 6.$$

Subtracting 3r from both sides, we obtain:

$$5r - 3r = 3r + 6 - 3r$$
$$2r = 6$$

or

$$r = 3.$$

If you were studying algebra, you would be working on problems of increasing complexity. The basic idea, however, would remain the same. That is, whatever you do to one side of the equation, you do to the other. If you follow that guideline, you can't go wrong.

APPLICATION TO PERCENT PROBLEMS

Suppose, at a clothing sale, everything has been marked down 40%. You find a suit you like for a sale price of $120. Let's try to figure out the original price.

Imagine yourself as the merchant. Marking the suit down 40% means the new price, $120, is 60% (100% – 40%) of the regular price. This can be written:

$$60\% \text{ of regular price} = \$120.$$

Since $60\% = \frac{60}{100} = \frac{3}{5}$ and "of" means multiply, we have:

$$\frac{3}{5} \times \text{regular price} = \$120.$$

This looks more like algebra if we have p stand for the regular price and write:

$$\frac{3}{5}p = 120.$$

Remember that a number ($\frac{3}{5}$) standing next to a letter means multiply. Can you solve this equation "intuitively"? If not, think about what to do to both sides of the equation so p, which you want to find, stands alone.

To solve the equation, multiply both sides by $\frac{5}{3}$. The motivation for this is that $\frac{5}{3} \times \frac{3}{5} = \frac{15}{15} = 1$, so you would end up with just one p:

$$\frac{5}{3} \times \frac{3}{5} p = \frac{5}{3} \times 120$$

$$p = \frac{600}{3}$$

$$p = 200$$

So the original price was $200. This checks because, if you work out 60% of 200, you get the sale price, $120.

How could you compute the dealer's cost if you knew the mark-up? For example, suppose you knew a stereo selling for $270 had been marked up 50% above wholesale.

Let's compute the wholesale price. First, think about how the retail price was obtained. The dealer added (marked up) 50% of cost to his original cost. This is summarized as:

Wholesale price + 50% of wholesale price = selling price

If p stands for wholesale price we have:

$$p + 50\% \text{ of } p = 270.$$

where $270 was the given selling price.

Since $50\% = \frac{1}{2}$ and "of" means multiply, we have:

$$p + \frac{1}{2}p = 270.$$

To simplify the left side, we note that one p plus an additional half p comes to one and a half p. Writing $1\frac{1}{2}$ as $\frac{3}{2}$ we have:

$$\frac{3}{2}p = 270.$$

To solve for p, multiply both sides of the equation by $\frac{2}{3}$ to obtain:

$$\frac{2}{3} \times \frac{3}{2}p = \frac{2}{3} \times 270$$

$$p = 180.$$

So the wholesale price was $180. This checks, because 180 plus 90 (50% of 180) is $270, the given selling price.

CHAPTER TWENTY-ONE

what is calculus anyway?

When you think of the word "calculus" what comes to mind? It sounds like a hard subject because the word is never used except in math. And you have heard friends who took it talking about having had difficulties.

As with any math subject, some concepts in calculus are difficult to grasp. Nevertheless, it is possible to develop an intuitive feeling for the nature of the subject without much math. That is the purpose of this chapter.

There are two parts to calculus, ultimately and surprisingly related to each other. Half of calculus has to do with finding the areas of funny shapes. By funny or unusual shapes we mean shapes, other than circles, that have curves in them. For example:

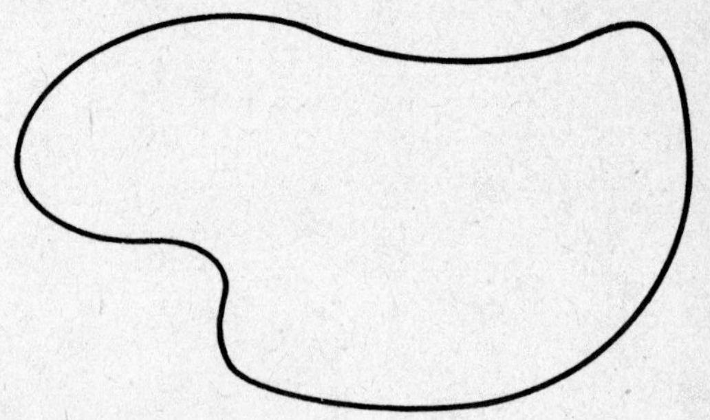

The hardest part of finding the area of a funny shape is in deciding where to start. A good guideline is to start with what you know best and then think about how that may help you.

The figures for which areas are easiest to calculate are rectangles. What is the area of this rectangle?

The answer is 8 square inches. Reviewing how that answer is obtained, we recall that a square inch is a square with each side measuring one inch. It looks like:

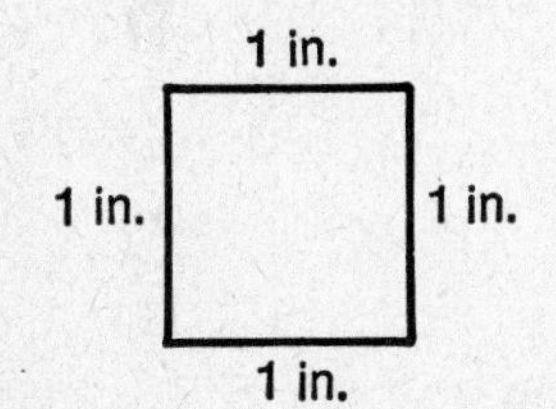

Asking for the area of the rectangle is asking how many square inches fit in it. (By the way, if the rectangle had been measured in feet or yards then the area would be in square feet or square yards.)

We mark off the above rectangle to illustrate that it contains 8 square inches.

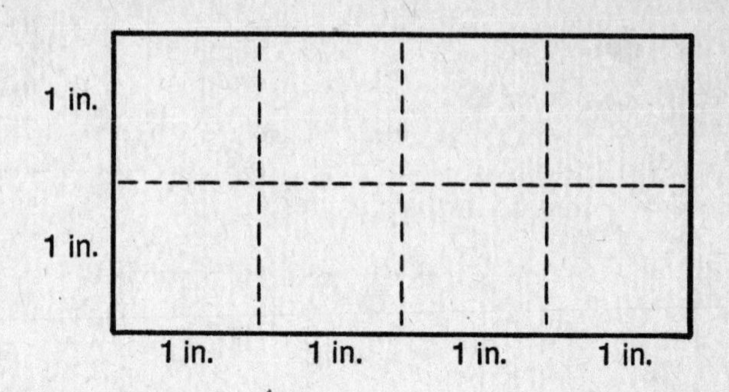

Summarizing, we say you can always get the area of a rectangle if you multiply length by width. The formula $A = LW$ was given in the last section.

Now we can go back to thinking about funny shapes. Let's start with a curved shape that is not too complicated looking, using the principle of starting with an easier rather than a harder problem. Concentrate on the shape:

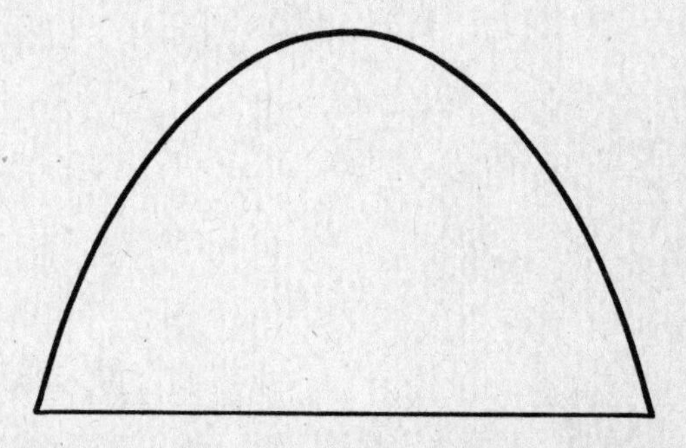

We will try to devise a method to approximate the area of this shape, using what we already know—rectangles. The first thing to do is draw a rectangle that fits inside the figure. You can do this however you like.

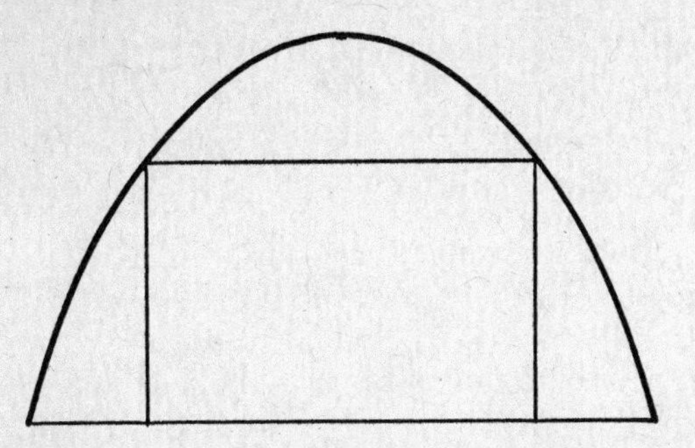

The area of the rectangle could be calculated if we measured it. This area gives an approximation to the total area—but not a very good one.

How could you make the approximation better? Could you add some more rectangles?

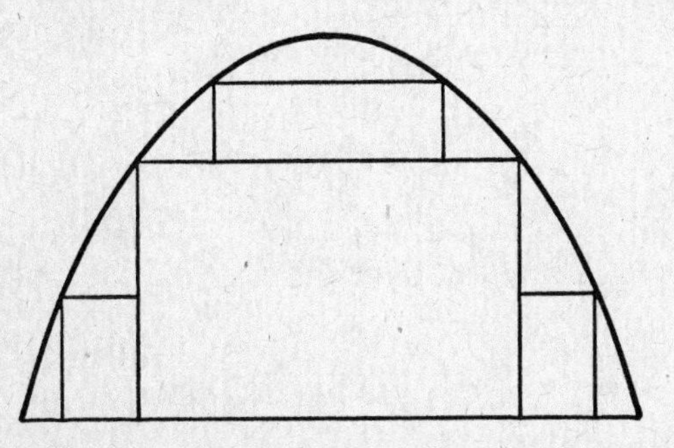

Don't be misled by some of the shapes outside the rectangles looking like triangles. With any curve, if you take a small section, it appears straight. Some parts of the outer edge of the figure may look almost straight, but in fact they have a very slight curve.

As before, we could measure the areas of all the rectangles

we have drawn. Adding those areas, we would get a better approximation to the area inside the figure.

We could then add still more rectangles to this figure and make the approximation even better. These rectangles would get tinier and tinier. Before doing this, however, we will shift our viewpoint to gain better insight into the pattern that is developing.

We extend the vertical sides of the top rectangle downward as dotted lines:

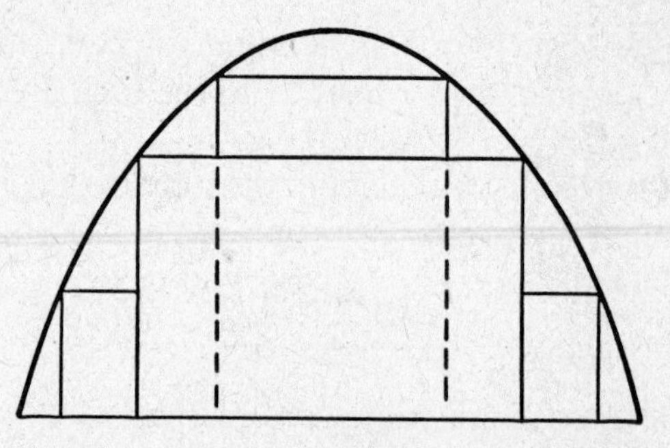

This has no effect on the combined area of the rectangles, but, after some thought, we make the drawing look like this:

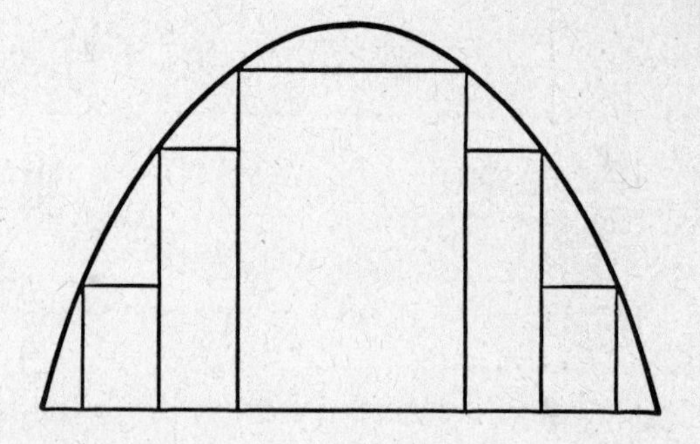

Compare the last two drawings to see that we have still covered the same area with rectangles. The difference is that the rectangles are now all vertical.

As we put in more and more vertical rectangles we get better and better approximations to the area under the curve:

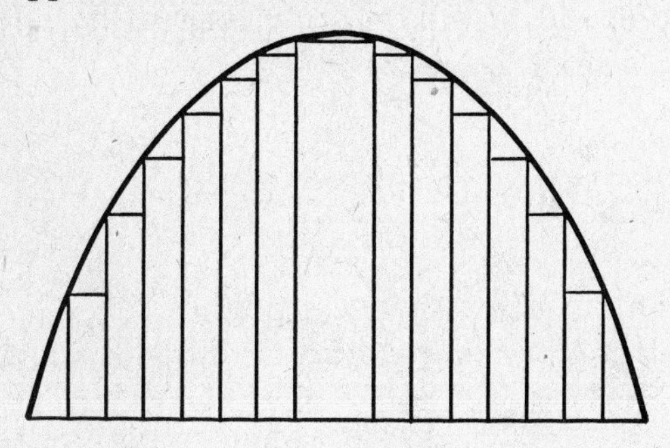

There is no end to this process and you might ask if it is ever possible to get the area exactly. But it looks as if we can get close enough. Does it seem that it would take infinitely many rectangles to fill up every last bit of area?

The concept of infinity is hard to grasp. This concept is part of the problem of finding the area under a curve. In a calculus course, weeks would be spent working with problems like the one just described.

Most remarkably, methods have been discovered which enable you to calculate the areas of curved figures exactly—without drawing rectangles and without doing much computation. Working with rectangles, as we have done, is the first step in the development of such methods.

To calculate the area of a "funny shape," you would have to

know the exact shape. In the terminology of calculus, curved shapes are referred to as "functions" and are studied extensively before trying to find their areas. The idea of infinity is dealt with through the study of another concept termed "limits." These terms refer to areas of conceptual difficulty in calculus. But the actual methods for calculating areas of curved shapes are not hard to learn.

MOOD SWINGS

We will begin our discussion of the other half of calculus by diagramming daily mood patterns. Think about how you feel when you first wake up. Do you wake right up or do you very gradually come awake during the course of the morning? When do you feel most alert? When do you feel worst?

We will draw a typical pattern—then explain it.

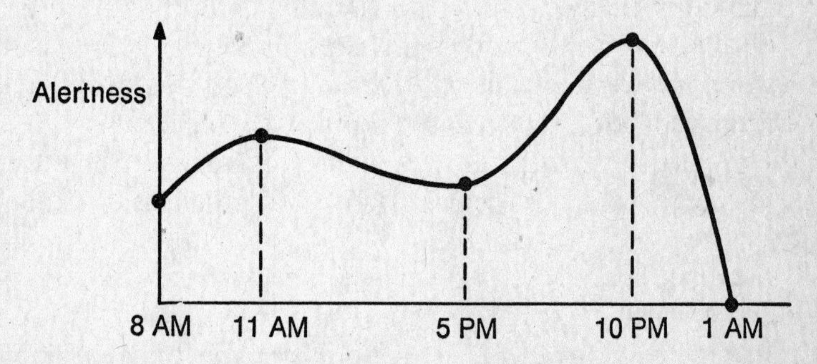

At 8:00 A.M. you wake up and are not very alert. During the morning, you feel better little by little, with 11:00 A.M. being the time you feel best. Then you gradually get more tired and by 5:00 P.M. you want to take a nap. Once you get past that time

you may start feeling better again. In fact, if you are a night person you may peak sometime later in the evening, at 10:00 P.M., for example, before getting tired again and falling asleep.

The diagram we have drawn has two highs (maxima) and one low (minimum). The second half of calculus is concerned with finding maxima and minima.

The graph shows a pattern of rising and falling between maximum and minimum points. Sometimes it changes slowly, as between 8 A.M. and 11 A.M., while at other times it changes very rapidly, as between 10 P.M. and 1 A.M. This half of calculus also studies rates of change, and this study has a wide variety of applications, from rockets (how fast they move and where they go) to the economy (when it is improving and when it is declining).

At the start of this chapter we mentioned that the two parts of calculus are related. This relationship is based on studying the rate of change of areas of curved shapes as the shapes themselves change.

We have highlighted the basic ideas of calculus. Filling in the details means taking a calculus course. But this should no longer be so mysterious or intimidating.